Microbe Hunting

Unveiling the Secrets of Microorganisms through Assessment, Sequencing, and Bioinformatics Analysis

Bhagwan Narayan Rekadwad

Division of Microbiology and Biotechnology
Yenepoya Research Centre
Yenepoya (Deemed to be University)
Mangalore, Karnataka, India

CRC Press
Taylor & Francis Group
Boca Raton London New York

CRC Press is an imprint of the
Taylor & Francis Group, an **informa** business

A SCIENCE PUBLISHERS BOOK

Cover credit: Image provided by the author.

First edition published 2025
by CRC Press
2385 NW Executive Center Drive, Suite 320, Boca Raton FL 33431

and by CRC Press
4 Park Square, Milton Park, Abingdon, Oxon, OX14 4RN

Library of Congress Cataloging-in-Publication Data (applied for)

ISBN: 978-1-032-75437-6 (hbk)
ISBN: 978-1-032-79471-6 (pbk)
ISBN: 978-1-003-49222-1 (ebk)

DOI: 10.1201/9781003492221

Typeset in Palatino Linotype
by Prime Publishing Services

Preface

Welcome, intrepid explorer, to the thrilling realm of microbiology, where invisible life forms hold the keys to a hidden universe. This book, Microbe Hunting: Unveiling the Secrets of Microorganisms through Assessment, Sequencing, and Bioinformatics Analysis, serves as your compass and guide on this captivating quest. Prepare to delve into the fascinating world of microbes, those microscopic beings that permeate every corner of our planet, from the deepest ocean trenches to the highest mountain peaks, and even within our own bodies. This book gives a base-line for Microbe Hunting through an understanding of Microbial Diversity and Ecosystems an advanced tool for huge and sophisticated analyses, and makes you delve into the future horizons of microbe hunting to shape the 22nd century and beyond. This book also includes a perspective on the use of quantum supercomputers, machine learning, deep learning, and artificial intelligence for "Microbe Hunting".

Throughout this book, every reader will encounter real-world examples, cutting-edge research, and practical guidance. Each chapter is designed to build your knowledge and equip you with the skills to become a skilled microbe hunter, ready to explore the hidden universe teeming around you. So, prepare to embark on this captivating adventure, unlock the secrets of the microbial world, and discover the profound impact these tiny life forms have on our planet and ourselves. Remember, the world of microbes is vast and ever-evolving. This book serves as a starting point and a springboard for your own exploration. Embrace the spirit of discovery, ask questions, delve deeper, and be prepared to be amazed by the hidden wonders revealed by your microbial hunt. The future of microbiology is bright, and you, the intrepid microbe hunter, hold the key to unlocking its secrets.

February 2024   Dr. Bhagwan Narayan Rekadwad

Contents

Chapter 1
Introduction to Microbe Hunting

Introduction to microbes

What are microbes?

Microbes are extremely minute living entities that are all around us and are invisible to the unaided eye. They are aquatic, terrestrial, and avian organisms. Millions of these bacteria, which are also referred to as microorganisms, reside in the human body (InformedHealth.org; Moënne-Loccoz et al., 2014). While certain microorganisms make us ill, others are vital to our wellbeing. The most prevalent kinds include fungus, viruses, and bacteria. Protozoa are a class of microorganisms as well. These are tiny living creatures that cause toxoplasmosis and malaria, among other illnesses (InformedHealth.org; Gerba, 2015; Maderspacher, 2020).

These unicellular bacteria can live in the presence oxygen, but others do not. Some people like it hot, while others like it cold (Hentges, 1996; Martin and Koonin, 2006; Souza, 2012; Fierer et al., 2017). *Salmonella* and *Staphylococcus* bacteria are two well-known types of bacteria. Most germs do not make people sick. Many of them even live on or in our bodies and help us stay healthy (Akbar and Anal, 2013; Kurtz et al., 2017; Rortana et al., 2021). For example, lactic acid bacteria in the intestines help us break down food. Some bacteria help the immune system get rid of germs by fighting them (Perdigón et al., 2001; Zhang et al., 2015). Some bacteria are also needed to make yogurt, kimchi, and cheese, among other things. Only a small percentage of bacteria cause diseases, but there are no exact numbers (Doron and Gorbach, 2008; Kok and Hutkins, 2018; Nagaoka, 2019). For example, only the pathogenic strain of *Mycobacterium tuberculosis* can cause tuberculosis (TB) (Chai et al., 2018; Bloom et al., 2017; Smith, 2003; Miggiano et al., 2020; Delogu et al., 2013; Rahlwes et al., 2023).

Antibiotics can be used to treat bacterial illnesses. These are medicines that kill germs or stop them from making more of themselves. Bacteria can also cause many other infections, like diarrhoea, colds, and tonsillitis (Antibiotics, 1999; Patel et al., 2024; Kotwani et al., 2021; Wall, 2019).

On other hand, viruses do not have their own cells like bacteria do. This means that, correctly speaking, they are not living things. Instead, they are made of one or more molecules with a protein shell around them. The viruses need the genetic information inside this shell in order to multiply. Many diseases are caused by viruses. Some are harmless and only cause a mild cold, while others can lead to dangerous diseases like AIDS (Fenner et al., 1987; Brüssow, 2009; Payne, 2017; Renneberg et al., 2017; Berman, 2019). Other diseases caused by viruses are influenza (also called "the flu"), measles, and viral hepatitis, which is an inflammation of the liver. Viruses get into healthy cells and use them to make more viruses (Drexler, 2010; O'Horo and Cawcutt, 2019; Thirumdas et al., 2021). Without these host cells, a virus cannot make copies of itself. Not all viruses cause signs, and most of the time, the body is able to fight back against the invaders. This is true of cold sores, which a lot of people have had at some point (Koyuncu et al., 2013; Ryu, 2017; Šudomová et al., 2022). They are caused by viruses that live in certain nerve cells. If a person's immune system is weak or worn out, they may get the usual blisters. Using medicine to fight bugs is not all that easy (National Research Council (US) Committee on Research Opportunities in Biology, 1989; Nicholson, 2016). The immune system can be "trained" to fight certain viruses by getting a vaccine. This makes the body better able to fight the virus (Informed Health, 2006; NIAID, 2006).

Types of microbes

-based on nature, size and shape

Microorganisms, which are also referred to as bacteria, are extremely small living organisms that can have a single cell, several cells, or clusters of cells (Figure 1). In the natural world, microorganisms may be found everywhere, and the majority of them are beneficial to life (Donlan, 2002; Gupta et al., 2016; Flint, 2020). But there are several of them that can cause significant harm. As far as microorganisms are concerned, there are six primary categories: bacteria, archaea, fungi, protozoa, algae, and viruses/prions (Refer: https://blog.foodsafety.ca/6-types-microorganisms-cause-food-borne-illness).

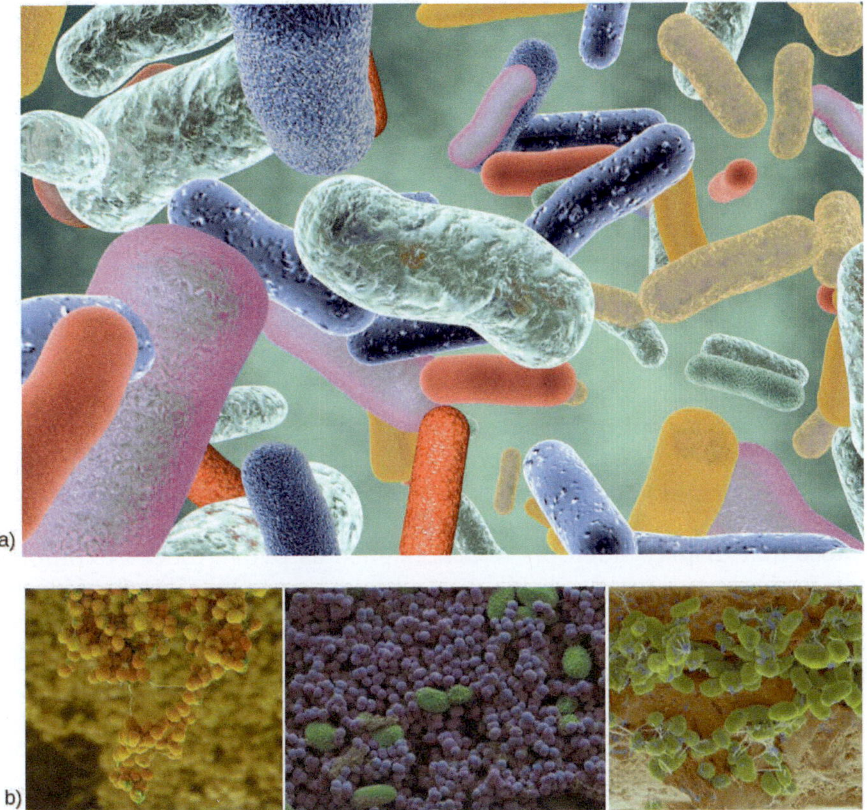

Figure 1. Microooorganisms: (a) Bacterial cell (Source: Science, https://www.science.org/content/resource/hunt-gut-microbe-molecule); (b) Different bacterial cells (Source: National Geographic Society, https://www.nationalgeographic.org/).

-based on cultivability (cultivable/uncultivable)

The paradigm whereby only 1% of microbes can be cultured (Stewart, 2012; Martiny, 2019) has had a profound effect on our comprehension of microbial ecology and remains a primary reason for relying primarily on molecular techniques for describing microbial communities (Robinson et al., 2010; Boughner and Singh, 2016). Nevertheless, this concept is frequently expressed in a nebulous manner, suggesting that certain researchers have divergent paradigmatic interpretations. In addition, significant advancements in cultivation techniques indicate that this paradigm may no longer be valid. To figure out how easy it is to grow bacteria in six main biomes, I found that the median 16S rRNA homology between bacteria and their known cultured relatives was 97.3 ± 2.3% (sd).

In addition, $52 \pm 24\%$ of sequences and $34.9 \pm 23\%$ of taxa (defined as > 97% similar) had a cultured relative that was closely related (Martiny, 2019).

The great majority of microbial species have not been "cultured" and hence do not grow when placed in a controlled environment such as a laboratory. Due to this, numerous techniques for cultivating these creatures in an atmosphere that is a reliable representation of their natural habitat have been developed. Methods include growing the organism in the presence of cultivable auxiliary species and placing the cells in chambers that permit the passage of substances from the natural environment (Prakash et al., 2020; Aguilera et al., 2021; Kapinusova et al., 2023). Other methods include traps encased with porous membranes that particularly capture organisms that generate hyphae, such as actinobacteria and microfungi (Rämä and Quandt, 2021; Gavrish et al., 2008; Lewis et al., 2010). Repeated culture *in situ* results in the generation of domesticated varieties, which are capable of growing on standard conditions *in vitro* and might be scaled up for the synthesis of secondary metabolites (Lewis et al., 2010). It was discovered that a number of different chemicals, including those that display siderophore action, such as pyoverdines-Fe-complex, desferricoprogen, and salicylic acid, might boost the growth of strains of bacteria that are notoriously difficult to culture, such as *Prevotella* sp. HOT-376 and *Fretibacterium fastidiosum* (Vartoukian et al., 2016). Therefore, the culture of uncultivated bacteria is nothing more than limitations of existing laboratory procedures rather than the ability of the bacteria to grow on a variety of laboratory media and under a variety of laboratory circumstances.

Importance of microbes

-Are we misunderstood microorganisms?

The vast majority of animals, including humans, engage in a wide range of interactions with microbes. Skin, hair, tongue, and stomach are all components that contribute to what is known as the "human microbiome." People have a tendency to associate disease with germs. However, the vast majority of the microbiome that we humans possess is either beneficial or neutral (Kennedy and Chang, 2020; Hollingsworth et al., 2021; Roy et al., 2022; Hou et al., 2022). The microbes that live in the intestines contribute to digestion, the absorption of nutrients, and the fight against germs that are harmful to the body. On the other hand, they are capable of producing proteins and vitamins that human genes are unable to make (Altveş et al., 2020; Wan et al., 2022). In addition to assisting the immune system in its fight against infections and diseases throughout the body, they also prevent the growth of germs that are harmful to the delicate skin. There are several scientific facts that can be used to gain an understanding of microorganisms (Canny et al., 2008; Morowitz et al., 2011; Belkaid and Hand, 2014; Jandhyala et al., 2015; Takiishi et al., 2017; Rowland et al., 2018; Rodríguez-Romero et al., 2021; Santhiravel et al., 2022).

Fact No. 1

The human cells outnumber the bacteria inside one person ten to one, and they can weigh up to three pounds.

Fact No. 2

Microbes also inhabit plants, soil, oceans, and the atmosphere in addition to the human microbiome. These microbiomes support the healthy operation of these different ecosystems, having an impact on things like food security, climate change, and human health.

Fact No. 3

One of the bacteria in our human microbiome that has been the subject of the most research is bacterium *Escherichia coli*. Although *E. coli* can cause serious intestinal diseases, it also contributes significantly to human ***vitamin K*** supply.

Fact No. 4

The air surrounding you can have a unique fingerprint-like quality. The microorganisms that the human body releases into the atmosphere can be utilized to uniquely identify the person.
(Source of information: Misunderstood Microbes, National Geographic)

Microbe hunting techniques

Sampling, isolation, and visualization

Microbiologists, sometimes famous as "microbe hunters," employ specimens or samples to separate and cultivate prokaryotic (bacteria, archaea), eukaryotic (fungi, animals), or viral species from collected samples. When bacteria are cultured *in vitro* or *in vivo*, the results can be considerably influenced by the isolation process. There are some species of bacteria that cannot always be separated, despite the fact that microbial isolation is a rather basic process (Hibbing et al., 2010; Ahmed, 2014; Shi et al., 2019). In situations where the sequencing of the pathogen candidate is already known, genomic reconstruction can serve as a feasible alternative to isolation. Although it is not possible to isolate, multiply, or reassemble a microbial cell, it is possible to visually examine it via the use of light or electron microscopy. Additionally, its proteins or nucleic acids can be examined through the use of immunohistochemistry or *in situ* hybridization. Interactions between antibodies and some potential substances can occur. Immunohistochemistry allows for the identification and verification of the presence of an agent as well as its distribution throughout the body (Arrigo and Lipkin, 2012).

Identification methods

Microorganism identification and characterization are essential components of research used to track out contaminants and solve issues like rotting. Knowing how to recognize bacteria or other unfamiliar microorganisms can help you determine whether they provide a risk for contamination or safety, or whether they are likely to withstand high temperatures (Jung et al., 2014; Lorenzo et al., 2018; Alegbeleye et al., 2018; Pauter et al., 2020; Mejia et al., 2022). However, there are situations when you need to know more, such as the source of a contamination. The only method to pinpoint the origins of a certain contaminant is to characterize it to a specific strain, compare it to a strain found elsewhere in the environment or in a specific food ingredient. As many organisms occupy a wide variety of habitats, this is the only way to do so (Carlin, 2011; Bergwerff and Debast, 2021; Lou et al., 2023; Santiago-Rodriguez and Hollister, 2023). Utilizing cutting-edge techniques, such as DNA sequencing, biochemical and phenotypic analysis, and identification of bacteria, moulds, and yeasts. Rapid pathogen confirmation by polymerase chain reaction, Repeat-based polymerase chain reaction, and riboprinter analysis are all methods for identifying and characterizing bacteria. Methods based on biochemistry, which can include a number of different methodologies and tests based on enzymes. PCR, or polymerase chain reaction, is a method of DNA "amplification" that can be used to identify specific genes or other DNA sequences (Anderson, 1994; Yang et al., 2008). This can be done, for instance, to verify the existence of particular microorganisms. Rep PCR is a type of PCR that detects' repeated DNA sequences' to create a DNA fragment pattern that helps characterize microorganisms. This pattern can be used to distinguish between different types of microorganisms (Healy et al., 2005; Spigaglia and Mastrantonio, 2003; Shin et al., 2023). Ribotyping is a technique that may be used for identification of microorganisms. It can categorize isolates according to the similarities of their DNA fragment pattern (characterization), or it can identify matches in patterns against a database of isolates (identification) (Jordens, 1998; Clermont et al., 2001; Register et al., 2003; Bouchet et al., 2008). Both of these functions are performed by the technique. The term "sequencing" refers to the process of analyzing the "DNA code" of a gene that is present in all bacteria or a gene that is present in all yeasts. This allows for a comparison to be made against a database of microorganisms to choose the organisms that are the most similar (Comden BRI, 2023).

Applications of microbes

In food quality and food safety

Microbes like bacteria, moulds, and yeasts have long been used in the food industry to make things like wine, beer, bread, and dairy products. Food production, preservation, and advancement all depend largely on microorganisms (Lorenzo et al., 2018; Raveendran et al., 2018; Mannaa et al., 2021). Functional microorganisms, especially yeast and bacteria, can increase the bioactivity of nutrients, produce antioxidant and antibacterial chemicals, and improve food safety. They can also enhance the sensory quality of food. Particularly in fermented foods, the type of microorganisms is tightly correlated with the physicochemical characteristics and volatile flavour components of food. When making salami sausage, Liu et al., used yeast and *Lactobacillus rhamnosus* YL-1 (Russo et al., 2007; Petrova et al., 2021) as a starter, which significantly reduced the level of lipid oxidation and led to alterations in the flavour profiles. *Staphylococcus saprophyticus* CGMCC 3475 (Ehlers and Merrill, 2023; Liu et al., 2023), leucine, and *Lactobacillus fermentum* YZU-06 (Naghmouchi et al., 2020; Liu et al., 2023) have all been used to produce fermented sausage, which raises the overall quality of the sausages as well as the variety of taste components. The chemical components of instant dark teas were altered by the fungi, as evidenced by the quality and primary metabolic alterations of instant dark teas fermented by various fungi, including *Aspergillus cristatus, Aspergillus niger,* and *Aspergillus tubingensis.* Using microbiome during fermentation could eliminate off-putting beany flavours and improve the olfactory profile of plant-based meat substitutes (Assad et al., 2023; Zhu et al., 2020; Liao et al., 2023). Probiotics significantly improve digestion and increase nutrient absorption, which both benefit gut health. Several variables, including temperature, time, pH, oxygen levels, and microbial starting cultures, have an impact on the fermentation process. By controlling the conditions of fermentation, food quality can be kept under observation. Achieving the desired qualities, producing dependable food products, and assuring food quality, safety, and consistency need the prudent application of helpful microbes and the control of harmful microbial infection (Cao et al., 2023).

Environmental monitoring and remediation

Microbes can be used to remediate the environment while monitoring it. Basically, there are three methods to go about it: physically, chemically, and biologically. Skimmers, sorbents, and booms are used in the physical remediation. Before a more extensive remediation method is carried out, a boom is a physical barrier consisting of materials that absorbs oil pollutants and keeps them from spreading. After booms, technologies like sorbents and skimmers are employed to further absorb and adsorb contaminants. As the bloom remediation technique depends on buoyancy and roll reaction, it presents the biggest problem. Booms that are buoyant float and stay on the water's surface for a longer period of time. The torque needed to rotate the bloom from its vertical position is referred to as the roll response. In other words, a higher roll response causes a higher remedial process. Chemical remediation is the process of stabilizing and removing heavy metals from the environment by introducing chemicals such clay minerals, phosphate, charcoal, aluminium salts, silicocalcium compounds, and sulphide. Adsorption, reduction, oxidation, complexation, precipitation, and ion exchange are some of the processes that underlie the utilization of these compounds. The use of chemicals in remediation is a quick, straightforward, and efficient method; nevertheless, the chemicals can also pollute the environment. Another way to treat pollution is by bioremediation, which is a long-term, low-cost, and secure procedure. The method uses organic materials like plants and microorganisms. Depending on the type, location, and degree of pollution, this approach may or may not be practical. On the other hand, microbes have shown to be effective at cleaning up environmental toxins. They are preferred to plants in remediation because they grow easily, have a quick growth cycle, and are simple to manipulate. To support a sustainable ecosystem, it is vital to improve the utilization of microorganisms as a bioremediation agent. They involves eight main mechanisms for microbial remediation such as (i) enzymatic oxidation, (ii) enzymatic reduction, (iii) bioaugmentation, (iv) biostimulation, (v) bioleaching, (vi) biosorption, (vii) bioaccumulation, and (viii) precipitation (Ayilara and Babalola, 2023).

Medicine and diagnostics

Microbes, including bacteria, viruses, fungi, and other microorganisms, have numerous applications in the field of medical and diagnostics. These applications leverage the unique characteristics and interactions of microbes to diagnose and treat various diseases. Here are some key applications of microbes in medical and diagnostic contexts:

1. *Microbial Cultures for Disease Diagnosis*: Microbes are grown in culture media to isolate and identify pathogens responsible for various infections. Bacterial cultures, for example, are used to identify the specific strain causing an infection, allowing for targeted treatment (Lagier et al., 2015).

2. *Antibiotic Production*: Certain microbes, such as *Penicillium notatum*, are used to produce antibiotics. These antibiotics are vital in treating bacterial infections and have saved countless lives (Gaynes, 2017).

3. *Vaccine Development*: Microbes are used to create vaccines by attenuating or inactivating pathogens. These vaccines stimulate the immune system to produce antibodies without causing illness, leading to immunity against specific diseases (Clem, 2011).

4. *Molecular Diagnostics*: Techniques like PCR (Polymerase Chain Reaction) and DNA sequencing rely on microbial enzymes (e.g., DNA polymerases) to amplify and analyze DNA and RNA, allowing for the detection of viruses, bacteria, and genetic mutations associated with diseases (Maheaswari et al., 2016; Dwivedi et al., 2017).

5. *Probiotics*: Beneficial microbes, such as various strains of bacteria and yeast, are used as probiotics to promote gut health and treat conditions like irritable bowel syndrome and diarrhoea (Kechagia et al., 2013; Amara and Shibl, 2015).

6. *Fermentation*: Microbes are employed in fermentation processes to produce medicines and biologics. For instance, insulin can be produced using genetically modified bacteria like *Escherichia coli* (Jozala et al., 2016; Pham et al., 2019).

7. *Bioremediation*: Microbes are used to clean up environmental contaminants, which indirectly benefit human health by reducing exposure to harmful substances (Bala et al., 2022).

8. *Diagnosis of Infectious Diseases*: Microbial assays are commonly used to diagnose infectious diseases. For example, ELISA (Enzyme-Linked Immunosorbent Assay) tests use microbial enzymes to detect antibodies or antigens associated with infections like HIV and hepatitis (Alhajj et al., 2023).

9. *Microbial Sensing and Detection*: Microbes can be engineered to act as biosensors for detecting specific molecules or pathogens. These biosensors can be used for rapid and sensitive diagnostics (Castillo-Henríquez et al., 2020).

10. *Phage Therapy*: Bacteriophages, which are viruses that infect bacteria, are being explored as a potential alternative to antibiotics to treat bacterial infections, particularly antibiotic-resistant strains (Lin et al., 2017).

11. *Microbiome Analysis*: Studying the human microbiome, which consists of trillions of microbes in and on the body, can provide insights into various health conditions, including gastrointestinal disorders, autoimmune diseases, and mental health disorders (Bull and Plummer, 2014; Ogunrinola et al., 2020).

12. *Gene Editing Tools*: Microbes like CRISPR-Cas9, a bacterial defence mechanism against viruses, have been adapted for precise gene editing in the treatment of genetic diseases and cancer (Li et al., 2020; Asmamaw and Zawdie, 2021).

13. *Biological Control*: Beneficial microbes are used to control disease-causing organisms in agriculture, protecting crops and reducing the need for chemical pesticides (He et al., 2021).

14. *Bacterial Biofilms*: Understanding microbial biofilms has applications in medical device development and infection prevention strategies (Wi and Patel, 2018; Khatoon et al., 2018).

15. *Microbial Forensics*: Microbial genetic analysis can be used in forensic investigations to trace the source of outbreaks or bioterrorism incidents (Committee on Science Needs for Microbial Forensics, 2014; Schmedes and Budowle, 2019).

16. *Microbial Biotechnology*: Microbes are utilized in the production of biopharmaceuticals, enzymes, and various medical products (Jozala et al., 2016).

These applications demonstrate the diverse roles that microbes play in medicine and diagnostics, from disease diagnosis and treatment to biotechnological advancements and environmental health. Researchers continue to explore new ways to harness the capabilities of microorganisms to improve human health and well-being.

Future of microbe hunting

Scientists and medical professionals have worked to comprehend and limit the transmission of infectious microorganisms ever since it was first realized that they are the root of many human diseases. These "microbe hunters" have mastered the art of fusing knowledge of the infectious agent(s) with epidemiological data over the course of the past 150+ years. Microbe hunting has changed across time, starting with the development of pre-germ theory epidemiological investigations, and continuing through the microbiological and molecular periods. In the genomic era, modern-day microbe hunters are combining epidemiological data with pathogen whole-genome sequencing to improve epidemiological investigations, advance our knowledge of the natural history of pathogens and disease drivers, and ultimately reshape our plans and priorities for worldwide disease control and eradication. Indeed, the work of microbe hunters is now more crucial than ever, as we have witnessed during the continuing Covid-19 outbreak. Microbial genomic epidemiology has come a long way, but there are still many interesting advancements in the works as the subject matures.

References

Aguilera, Á., de Diego-Castilla, G., Osuna, S., Bardera, R., Sor Mendi, S., Blanco, Y. et al. (2021). Microbial ecology in the atmosphere: the last extreme environment. IntechOpen. doi: 10.5772/intechopen.81650.

Ahmed, N. H. (2014). Cultivation of parasites. Tropical Parasitology 4(2): 80–89. https://doi.org/10.4103/2229-5070.138534.

Akbar, A. and Anal, A. K. (2013). Prevalence and antibiogram study of Salmonella and *Staphylococcus aureus* in poultry meat. Asian Pacific Journal of Tropical Biomedicine 3(2): 163–168. https://doi.org/10.1016/S2221-1691(13)60043-X.

Alegbeleye, O. O., Singleton, I. and Sant'Ana, A. S. (2018). Sources and contamination routes of microbial pathogens to fresh produce during field cultivation: A review. Food Microbiology 73: 177–208. https://doi.org/10.1016/j.fm.2018.01.003.

Alhajj, M., Zubair, M. and Farhana, A. (2023). Enzyme linked immunosorbent assay. [Updated 2023 Apr 23]. *In*: StatPearls [Internet]. Treasure Island (FL): StatPearls Publishing; 2023 Jan-. Available from: https://www.ncbi.nlm.nih.gov/books/NBK555922/.

Altveş, S., Yildiz, H. K. and Vural, H. C. (2020). Interaction of the microbiota with the human body in health and diseases. Bioscience of Microbiota, Food and Health 39(2): 23–32. https://doi.org/10.12938/bmfh.19-023.

Amara, A. A. and Shibl, A. (2015). Role of probiotics in health improvement, infection control and disease treatment and management. Saudi Pharmaceutical Journal: SPJ: The Official Publication of the Saudi Pharmaceutical Society 23(2): 107–114. https://doi.org/10.1016/j.jsps.2013.07.001.

Anderson, B. (1994). Broad-range polymerase chain reaction for detection and identification of bacteria. The Journal of the Florida Medical Association 81(12): 835–837.

Antibiotics. (1999). Paediatrics & Child Health 4(7): 504.

Arrigo, N. C. and Lipkin W. I. (2012). Microbe hunting and pathogen discovery. *In*: Institute of Medicine (US). Improving Food Safety Through a One Health Approach: Workshop Summary. Washington (DC): National Academies Press (US). A10. Available from: https://www.ncbi.nlm.nih.gov/books/NBK114503/.

Asmamaw, M. and Zawdie, B. (2021). Mechanism and applications of CRISPR/Cas-9-mediated genome editing. Biologics: Targets & Therapy 15: 353–361. https://doi.org/10.2147/BTT.S326422.

Assad, M., Ashaolu, T. J., Khalifa, I., Baky, M. H. and Farag, M. A. (2023). Dissecting the role of microorganisms in tea production of different fermentation levels: a multifaceted review of their action mechanisms, quality attributes and future perspectives. World Journal of Microbiology & Biotechnology 39(10): 265. https://doi.org/10.1007/s11274-023-03701-5.

Ayilara, M. S. and Babalola, O. O. (2023). Bioremediation of environmental wastes: the role of microorganisms. Front Agron 5: 1183691. doi: 10.3389/fagro.2023.1183691.

Baker, K. S. (2020). Microbe hunting in the modern era: reflecting on a decade of microbial genomic epidemiology. Current Biology: CB 30(19): R1124–R1130. https://doi.org/10.1016/j.cub.2020.06.097.

Bala, S., Garg, D., Thirumalesh, B. V., Sharma, M., Sridhar, K., Inbaraj, B. S. et al. 2022. Recent strategies for bioremediation of emerging pollutants: a review for a green and sustainable environment. Toxics 10(8): 484. https://doi.org/10.3390/toxics10080484.

Belkaid, Y. and Hand, T. W. (2014). Role of the microbiota in immunity and inflammation. Cell 157(1): 121–141. https://doi.org/10.1016/j.cell.2014.03.011.

Bergwerff, A. A. and Debast, S. B. (2021). Modernization of control of pathogenic micro-organisms in the food-chain requires a durable role for immunoaffinity-based detection methodology—a review. Foods (Basel, Switzerland) 10(4): 832. https://doi.org/10.3390/foods10040832.

Berman, J. J. (2019). Viruses. Taxonomic Guide to Infectious Diseases, 263–319. https://doi.org/10.1016/B978-0-12-817576-7.00007-9.

Bloom, B. R., Atun, R., Cohen, T., Dye, C., Fraser, H., Gomez, G. B. et al. (2017). Tuberculosis. *In*: Holmes, K. K. et al. (eds.). Major Infectious Diseases. (3rd ed.). The International Bank for Reconstruction and Development/The World Bank.

Bouchet, V., Huot, H. and Goldstein, R. (2008). Molecular genetic basis of ribotyping. Clinical Microbiology Reviews 21(2): 262–273. https://doi.org/10.1128/CMR.00026-07.

Boughner, L. A. and Singh, P. (2016). Microbial Ecology: Where are we now? Postdoc Journal: A Journal of Postdoctoral Research and Postdoctoral Affairs 4(11): 3–17. https://doi.org/10.14304/SURYA.JPR.V4N11.2.

Brüssow, H. (2009). The not so universal tree of life or the place of viruses in the living world. Philosophical Transactions of the Royal Society of London. Series B, Biological Sciences 364(1527): 2263–2274. https://doi.org/10.1098/rstb.2009.0036.

Bull, M. J. and Plummer, N. T. (2014). Part 1: The Human Gut Microbiome in Health and Disease. Integrative medicine (Encinitas, Calif.) 13(6): 17–22.

Canny, G. O. and McCormick, B. A. (2008). Bacteria in the intestine, helpful residents or enemies from within? Infection and Immunity 76(8): 3360–3373. https://doi.org/10.1128/IAI.00187-08.

Cao, J., Wang, Y., Zhou, C. and Geng, F. (2023). Editorial: Insights into the role of microorganisms on food quality and food safety. Front Microbiol. 14: 1237508. doi: 10.3389/fmicb.2023.1237508.

Carlin, F. (2011). Origin of bacterial spores contaminating foods. Food Microbiology 28(2): 177–182. https://doi.org/10.1016/j.fm.2010.07.008.

Castillo-Henríquez, L., Brenes-Acuña, M., Castro-Rojas, A., Cordero-Salmerón, R., Lopretti-Correa, M. and Vega-Baudrit, J. R. (2020). Biosensors for the detection of bacterial and viral clinical pathogens. Sensors (Basel, Switzerland) 20(23): 6926. https://doi.org/10.3390/s20236926.

Chai, Q., Zhang, Y. and Liu, C. H. (2018). Mycobacterium tuberculosis: An adaptable pathogen associated with multiple human diseases. Frontiers in Cellular and Infection Microbiology 8: 158. https://doi.org/10.3389/fcimb.2018.00158.

Clem, A. S. (2011). Fundamentals of vaccine immunology. Journal of Global Infectious Diseases 3(1): 73–78. https://doi.org/10.4103/0974-777X.77299.

Clermont, O., Cordevant, C., Bonacorsi, S., Marecat, A., Lange, M. and Bingen, E. (2001). Automated ribotyping provides rapid phylogenetic subgroup affiliation of clinical extraintestinal pathogenic *Escherichia coli* strains. Journal of Clinical Microbiology 39(12): 4549–4553. https://doi.org/10.1128/JCM.39.12.4549-4553.2001.

Committee on Science Needs for Microbial Forensics: Developing an Initial International Roadmap; Board on Life Sciences; Division on Earth and Life Studies; National Research Council. Science Needs for Microbial Forensics: Developing Initial International Research Priorities. Washington (DC): National Academies Press (US); 2014 Jul 25. 1, Introduction: What Is Microbial Forensics and Why Is It Important. Available from: https://www.ncbi.nlm.nih.gov/books/NBK234883/.

Compden BRI. 2023. Microorganism identification and characterisation. https://www.campdenbri.co.uk/services/microorganism-identification.php Assessed on September 14, 2023.

Delogu, G., Sali, M. and Fadda, G. (2013). The biology of mycobacterium tuberculosis infection. Mediterranean Journal of Hematology and Infectious Diseases 5(1): e2013070. https://doi.org/10.4084/MJHID.2013.070.

Donlan, R. M. (2002). Biofilms: microbial life on surfaces. Emerging Infectious Diseases 8(9): 881–890. https://doi.org/10.3201/eid0809.020063.

Doron, S. and Gorbach, S. L. (2008). Bacterial infections: overview. International Encyclopedia of Public Health 273–282. https://doi.org/10.1016/B978-012373960-5.00596-7.

Drexler M; Institute of Medicine (US). What You Need to Know About Infectious Disease. Washington (DC): National Academies Press (US); 2010. II, Disease Threats. Available from: https://www.ncbi.nlm.nih.gov/books/NBK209711/.

Dwivedi, S., Purohit, P., Misra, R., Pareek, P., Goel, A., Khattri, S. et al. 2017. Diseases and molecular diagnostics: a step closer to precision medicine. Indian Journal of Clinical Biochemistry: IJCB 32(4): 374–398. https://doi.org/10.1007/s12291-017-0688-8.

Ehlers, S. and Merrill, S. A. (2023). Staphylococcus saprophyticus Infection. In StatPearls. StatPearls Publishing.

Fenner, F., Bachmann, P. A., Gibbs, E. P. J., Murphy, F. A., Studdert, M. J. and White, D. O. (1987). Structure and composition of viruses. Veterinary Virology 3–19. https://doi.org/10.1016/B978-0-12-253055-5.50005-0.

Fierer, J., Looney, D. and Pechère, J. C. (2017). Nature and pathogenicity of micro-organisms. Infectious Diseases 4–25.e1. https://doi.org/10.1016/B978-0-7020-6285-8.00002-2.

Flint, H. J. (2020). Micro-organisms and the microbiome. Why Gut Microbes Matter: Understanding Our Microbiome, 1–8. https://doi.org/10.1007/978-3-030-43246-1_1.

Gavrish, E., Bollmann, A., Epstein, S. and Lewis, K. (2008). A trap for *in situ* cultivation of filamentous actinobacteria. Journal of Microbiological Methods 72(3): 257–262. https://doi.org/10.1016/j.mimet.2007.12.009.

Gaynes, R. 2017. The discovery of penicillin—new insights after more than 75 years of clinical use. Emerging Infectious Diseases 23(5): 849–853. https://doi.org/10.3201/eid2305.161556.

Gerba, C. P. (2015). Environmentally transmitted pathogens. Environmental Microbiology 509–550. https://doi.org/10.1016/B978-0-12-394626-3.00022-3.

Gupta, A., Gupta, R. and Singh, R. L. (2016). Microbes and environment. Principles and Applications of Environmental Biotechnology for a Sustainable Future 43–84. https://doi.org/10.1007/978-981-10-1866-4_3.

He, D. C., He, M. H., Amalin, D. M., Liu, W., Alvindia, D. G. and Zhan, J. 2021. Biological control of plant diseases: an evolutionary and eco-economic consideration. Pathogens (Basel, Switzerland) 10(10): 1311. https://doi.org/10.3390/pathogens10101311.

Healy, M., Huong, J., Bittner, T., Lising, M., Frye, S., Raza, S. et al. (2005). Microbial DNA typing by automated repetitive-sequence-based PCR. Journal of Clinical Microbiology 43(1): 199–207. https://doi.org/10.1128/JCM.43.1.199-207.2005.

Hentges, D. J. (1996). Anaerobes: General characteristics. *In*: Baron, S. (ed.). Medical Microbiology. 4th edition. Galveston (TX): University of Texas Medical Branch at Galveston. Chapter 17. Available from: https://www.ncbi.nlm.nih.gov/books/NBK7638/.

Hibbing, M. E., Fuqua, C., Parsek, M. R. and Peterson, S. B. (2010). Bacterial competition: surviving and thriving in the microbial jungle. Nature Reviews. Microbiology 8(1): 15–25. https://doi.org/10.1038/nrmicro2259.

Hollingsworth, B. A., Cassatt, D. R., DiCarlo, A. L., Rios, C. I., Satyamitra, M. M., Winters, T. A. et al. (2021). Acute radiation syndrome and the microbiome: impact and review. Frontiers in Pharmacology 12: 643283. https://doi.org/10.3389/fphar.2021.643283.

Hou, K., Wu, Z. X., Chen, X. Y., Wang, J. Q., Zhang, D., Xiao, C. et al. (2022). Microbiota in health and diseases. Signal Transduction and Targeted Therapy 7(1): 135. https://doi.org/10.1038/s41392-022-00974-4.

InformedHealth.org [Internet]. Cologne, Germany: Institute for Quality and Efficiency in Health Care (IQWiG); 2006. What are microbes? 2010 Oct 6 [Updated 2019 Aug 29]. Available from: https://www.ncbi.nlm.nih.gov/books/NBK279387/.

Jandhyala, S. M., Talukdar, R., Subramanyam, C., Vuyyuru, H., Sasikala, M. and Nageshwar Reddy, D. (2015). Role of the normal gut microbiota. World Journal of Gastroenterology 21(29): 8787–8803. https://doi.org/10.3748/wjg.v21.i29.8787.

Jordens, J. Z. (1998). Genomic DNA digestion and ribotyping. Methods in Molecular Medicine 15: 17–31. https://doi.org/10.1385/0-89603-498-4:17.

Jozala, A. F., Geraldes, D. C., Tundisi, L. L., Feitosa, V. A., Breyer, C. A., Cardoso, S. L. et al. (2016). Biopharmaceuticals from microorganisms: from production to purification. Brazilian Journal of Microbiology: [publication of the Brazilian Society for Microbiology] 47 Suppl 1(Suppl 1): 51–63. https://doi.org/10.1016/j.bjm.2016.10.007.

Jung, A. V., Le Cann, P., Roig, B., Thomas, O., Baurès, E. and Thomas, M. F. (2014). Microbial contamination detection in water resources: interest of current optical methods, trends and needs in the context of climate change. International Journal of Environmental Research and Public Health 11(4): 4292–4310. https://doi.org/10.3390/ijerph110404292.

Kapinusova, G., Lopez Marin, M. A. and Uhlik, O. (2023). Reaching unreachables: Obstacles and successes of microbial cultivation and their reasons. Frontiers in Microbiology 14: 1089630. https://doi.org/10.3389/fmicb.2023.1089630.

Kechagia, M., Basoulis, D., Konstantopoulou, S., Dimitriadi, D., Gyftopoulou, K., Skarmoutsou, N. et al. (2013). Health benefits of probiotics: a review. ISRN Nutrition 2013: 481651. https://doi.org/10.5402/2013/481651.

Kennedy, M. S. and Chang, E. B. (2020). The microbiome: Composition and locations. Progress in Molecular Biology and Translational Science 176: 1–42. https://doi.org/10.1016/bs.pmbts.2020.08.013.

Khatoon, Z., McTiernan, C. D., Suuronen, E. J., Mah, T. F. and Alarcon, E. I. (2018). Bacterial biofilm formation on implantable devices and approaches to its treatment and prevention. Heliyon 4(12): e01067. https://doi.org/10.1016/j.heliyon.2018.e01067.

Kok, C. R. and Hutkins, R. (2018). Yogurt and other fermented foods as sources of health-promoting bacteria. Nutrition Reviews 76(Suppl 1): 4–15. https://doi.org/10.1093/nutrit/nuy056.

Kotwani, A., Joshi, J. and Lamkang, A. S. (2021). Over-the-counter sale of antibiotics in india: a qualitative study of providers' perspectives across two states. Antibiotics (Basel, Switzerland) 10(9): 1123. https://doi.org/10.3390/antibiotics10091123.

Koyuncu, O. O., Hogue, I. B. and Enquist, L. W. (2013). Virus infections in the nervous system. Cell Host & Microbe 13(4): 379–393. https://doi.org/10.1016/j.chom.2013.03.010.

Kurtz, J. R., Goggins, J. A. and McLachlan, J. B. (2017). Salmonella infection: Interplay between the bacteria and host immune system. Immunology Letters 190: 42–50. https://doi.org/10.1016/j.imlet.2017.07.006.

Lagier, J. C., Edouard, S., Pagnier, I., Mediannikov, O., Drancourt, M. and Raoult, D. (2015). Current and past strategies for bacterial culture in clinical microbiology. Clinical Microbiology Reviews 28(1): 208–236. https://doi.org/10.1128/CMR.00110-14.

Lewis, K., Epstein, S., D'Onofrio, A. and Ling, L. L. (2010). Uncultured microorganisms as a source of secondary metabolites. The Journal of Antibiotics 63(8): 468–476. https://doi.org/10.1038/ja.2010.87.

Lewis, K., Epstein, S., D'Onofrio, A. and Ling, L. (2010). Uncultured microorganisms as a source of secondary metabolites. J. Antibiot. 63: 468–476. https://doi.org/10.1038/ja.2010.87.

Li, H., Yang, Y., Hong, W., Huang, M., Wu, M. and Zhao, X. (2020). Applications of genome editing technology in the targeted therapy of human diseases: mechanisms, advances and prospects. Signal Transduction and Targeted Therapy 5(1). https://doi.org/10.1038/s41392-019-0089-y.

Liao, S. Y., Zhao, Y. Q., Jia, W. B., Niu, L., Bouphun, T., Li, P. W. et al. (2023). Untargeted metabolomics and quantification analysis reveal the shift of chemical constituents between instant dark teas individually liquid-state fermented by *Aspergillus cristatus*, *Aspergillus niger*, and *Aspergillus tubingensis*. Frontiers in Microbiology 14: 1124546. https://doi.org/10.3389/fmicb.2023.1124546.

Lin, D. M., Koskella, B. and Lin, H. C. 2017. Phage therapy: An alternative to antibiotics in the age of multi-drug resistance. World Journal of Gastrointestinal Pharmacology and Therapeutics 8(3): 162–173. https://doi.org/10.4292/wjgpt.v8.i3.162.

Liu, R., Ma, Y., Chen, L., Lu, C., Ge, Q., Wu, M. et al. (2023). Effects of the addition of leucine on flavor and quality of sausage fermented by *Lactobacillus fermentum* YZU-06 and *Staphylococcus saprophyticus* CGMCC 3475. Frontiers in Microbiology 13: 1118907. https://doi.org/10.3389/fmicb.2022.1118907.

Lorenzo, J. M., Munekata, P. E., Dominguez, R., Pateiro, M., Saraiva, J. A. and Franco, D. (2018). Main groups of microorganisms of relevance for food safety and stability: general aspects and overall description. Innovative Technologies for Food Preservation 53–107. https://doi.org/10.1016/B978-0-12-811031-7.00003-0.

Lou, Y. C., Hoff, J., Olm, M. R., West-Roberts, J., Diamond, S., Firek, B. A. et al. (2023). Using strain-resolved analysis to identify contamination in metagenomics data. Microbiome 11(1): 36. https://doi.org/10.1186/s40168-023-01477-2.

Maderspacher, F. (2020). Consider the microbe. Current Biology: CB 30(19): R1096–R1099. https://doi.org/10.1016/j.cub.2020.09.010.

Maheaswari, R., Kshirsagar, J. T. and Lavanya, N. 2016. Polymerase chain reaction: A molecular diagnostic tool in periodontology. Journal of Indian Society of Periodontology 20(2): 128–135. https://doi.org/10.4103/0972-124X.176391.

Mannaa, M., Han, G., Seo, Y. S. and Park, I. (2021). Evolution of food fermentation processes and the use of multi-omics in deciphering the roles of the microbiota. Foods (Basel, Switzerland) 10(11): 2861. https://doi.org/10.3390/foods10112861.

Martin, W. and Koonin, E. V. (2006). A positive definition of prokaryotes. Nature 442(7105): 868. https://doi.org/10.1038/442868c.

Martiny, A. C. (2019). High proportions of bacteria are culturable across major biomes. The ISME Journal 13(8): 2125–2128. https://doi.org/10.1038/s41396-019-0410-3.

Mejia, O. A. V. and Fernandes, P. M. P. (2022). Checklists as a central part of surgical safety culture. Sao Paulo Medical Journal = Revista paulista de medicina 140(4): 515–517. https://doi.org/10.1590/1516-3180.2022.140404052022.

Miggiano, R., Rizzi, M. and Ferraris, D. M. (2020). Mycobacterium tuberculosis pathogenesis, infection prevention and treatment. Pathogens (Basel, Switzerland) 9(5): 385. https://doi.org/10.3390/pathogens9050385.

Misunderstood Microbes, National Geographic. https://education.nationalgeographic.org/resource/misunderstood-microbes/ Assessed on September 11, 2023, Time 16:45:21.

Moënne-Loccoz, Y., Mavingui, P., Combes, C., Normand, P. and Steinberg, C. (2014). Microorganisms and biotic interactions. Environmental Microbiology: Fundamentals and Applications: Microbial Ecology 395–444. https://doi.org/10.1007/978-94-017-9118-2_11.

Morowitz, M. J., Carlisle, E. M. and Alverdy, J. C. (2011). Contributions of intestinal bacteria to nutrition and metabolism in the critically ill. The Surgical Clinics of North America 91(4): 771–viii. https://doi.org/10.1016/j.suc.2011.05.001.

Nagaoka, S. (2019). Yogurt Production. Methods in molecular biology (Clifton, N.J.) 1887: 45–54. https://doi.org/10.1007/978-1-4939-8907-2_5.

Naghmouchi, K., Belguesmia, Y., Bendali, F., Spano, G., Seal, B. S. and Drider, D. (2020). Lactobacillus fermentum: a bacterial species with potential for food preservation and biomedical applications. Critical Reviews in Food Science and Nutrition 60(20): 3387–3399. https://doi.org/10.1080/10408398.2019.1688250.

National Institute of Allergy and Infectious Diseases (NIAID). (2006). Understanding Microbes in Sickness and in Health. NIH Publication No. 09-4941.

National Research Council (US) Committee on Research Opportunities in Biology. Opportunities in Biology. Washington (DC): National Academies Press (US). (1989). 7, The Immune System and Infectious Diseases. Available from: https://www.ncbi.nlm.nih.gov/books/NBK217803/.

Nicholson, L. B. (2016). The immune system. Essays in Biochemistry 60(3): 275–301. https://doi.org/10.1042/EBC20160017.

O'Horo, J. C. and Cawcutt, K. A. (2019). Critical care viral infections. Critical Care Nephrology 560–567.e1. https://doi.org/10.1016/B978-0-323-44942-7.00096-0.

Ogunrinola, G. A., Oyewale, J. O., Oshamika, O. O. and Olasehinde, G. I. (2020). The Human microbiome and its impacts on health. International Journal of Microbiology 2020: 8045646. https://doi.org/10.1155/2020/8045646.

Patel, P., Wermuth, H. R., Calhoun, C. and Hall, G. A. (2023). Antibiotics. [Updated 2023 May 26]. *In*: StatPearls [Internet]. Treasure Island (FL): StatPearls Publishing; 2024 Jan-. Available from: https://www.ncbi.nlm.nih.gov/books/NBK535443/.

Pauter, K., Szultka-Młyńska, M. and Buszewski, B. (2020). Determination and identification of antibiotic drugs and bacterial strains in biological samples. Molecules (Basel, Switzerland) 25(11): 2556. https://doi.org/10.3390/molecules25112556.

Payne, S. (2017). Introduction to animal viruses. Viruses 1–11. https://doi.org/10.1016/B978-0-12-803109-4.00001-5.

Perdigón, G., Fuller, R. and Raya, R. (2001). Lactic acid bacteria and their effect on the immune system. Current Issues in Intestinal Microbiology 2(1): 27–42.

Petrova, M. I., Reid, G. and Ter Haar, J. A. (2021). Lacticaseibacillus rhamnosus GR-1, a.k.a. Lactobacillus rhamnosus GR-1: Past and future perspectives. Trends in Microbiology 29(8): 747–761. https://doi.org/10.1016/j.tim.2021.03.010.

Pham, J. V., Yilma, M. A., Feliz, A., Majid, M. T., Maffetone, N., Walker, J. R. et al. (2019). A review of the microbial production of bioactive natural products and biologics. Frontiers in Microbiology 10: 1404. https://doi.org/10.3389/fmicb.2019.01404.

Prakash, O., Nimonkar, Y. and Desai, D. (2020). A recent overview of microbes and microbiome preservation. Indian Journal of Microbiology 60(3): 297–309. https://doi.org/10.1007/s12088-020-00880-9.

Rahlwes, K. C., Dias, B. R. S., Campos, P. C., Alvarez-Arguedas, S. and Shiloh, M. U. (2023). Pathogenicity and virulence of Mycobacterium tuberculosis. Virulence 14(1): 2150449. https://doi.org/10.1080/21505594.2022.2150449.

Rämä, T. and Quandt, C. A. (2021). Improving fungal cultivability for natural products discovery. Frontiers in Microbiology 12: 706044. https://doi.org/10.3389/fmicb.2021.706044.

Raveendran, S., Parameswaran, B., Ummalyma, S. B., Abraham, A., Mathew, A. K., Madhavan, A. et al. (2018). Applications of microbial enzymes in food industry. Food Technology and Biotechnology 56(1): 16–30. https://doi.org/10.17113/ftb.56.01.18.5491.

Register, K. B., Sacco, R. E. and Nordholm, G. E. (2003). Comparison of ribotyping and restriction enzyme analysis for inter- and intraspecies discrimination of Bordetella avium and Bordetella hinzii. Journal of Clinical Microbiology 41(4): 1512–1519. https://doi.org/10.1128/JCM.41.4.1512-1519.2003.

Renneberg, R., Berkling, V. and Loroch, V. (2017). Viruses, antibodies, and vaccines. Biotechnology for Beginners 165–200. https://doi.org/10.1016/B978-0-12-801224-6.00005-9.

Robinson, C. J., Bohannan, B. J. and Young, V. B. (2010). From structure to function: the ecology of host-associated microbial communities. Microbiology and Molecular Biology Reviews: MMBR 74(3): 453–476. https://doi.org/10.1128/MMBR.00014-10.

Rodríguez-Romero, J. J., Durán-Castañeda, A. C., Cárdenas-Castro, A. P., Sánchez-Burgos, J. A., Zamora-Gasga, V. M. and Sáyago-Ayerdi, S. G. (2021). What we know about protein gut metabolites: Implications and insights for human health and diseases. Food Chemistry: X 13: 100195. https://doi.org/10.1016/j.fochx.2021.100195.

Rortana, C., Nguyen-Viet, H., Tum, S., Unger, F., Boqvist, S., Dang-Xuan, S. et al. (2021). Prevalence of Salmonella spp. and *Staphylococcus aureus* in Chicken Meat and Pork from Cambodian Markets. Pathogens (Basel, Switzerland) 10(5): 556. https://doi.org/10.3390/pathogens10050556.

Rowland, I., Gibson, G., Heinken, A., Scott, K., Swann, J., Thiele, I. et al. (2018). Gut microbiota functions: metabolism of nutrients and other food components. European Journal of Nutrition 57(1): 1–24. https://doi.org/10.1007/s00394-017-1445-8.

Roy, S., Nag, S., Saini, A. and Choudhury, L. (2022). Association of human gut microbiota with rare diseases: A close peep through. Intractable & Rare Diseases Research 11(2): 52–62. https://doi.org/10.5582/irdr.2022.01025.

Russo, F., Orlando, A., Linsalata, M., Cavallini, A. and Messa, C. (2007). Effects of *Lactobacillus rhamnosus* GG on the cell growth and polyamine metabolism in HGC-27 human gastric cancer cells. Nutrition and Cancer 59(1): 106–114. https://doi.org/10.1080/01635580701365084.

Ryu, W. S. (2017). Virus life cycle. Molecular Virology of Human Pathogenic Viruses 31–45. https://doi.org/10.1016/B978-0-12-800838-6.00003-5.

Santhiravel, S., Bekhit, A. E. A., Mendis, E., Jacobs, J. L., Dunshea, F. R., Rajapakse, N. et al. (2022). The impact of plant phytochemicals on the gut microbiota of humans for a balanced life. International Journal of Molecular Sciences 23(15): 8124. https://doi.org/10.3390/ijms23158124.

Santiago-Rodriguez, T. M. and Hollister, E. B. (2023). Viral metagenomics as a tool to track sources of fecal contamination: a one health approach. Viruses 15(1): 236. https://doi.org/10.3390/v15010236.

Schmedes, S. and Budowle, B. (2019). Microbial forensics. Encyclopedia of Microbiology 134–145. https://doi.org/10.1016/B978-0-12-801238-3.02483-1.

Shi, D., Mi, G., Wang, M. and Webster, T. J. (2019). *In vitro* and *ex vivo* systems at the forefront of infection modeling and drug discovery. Biomaterials 198: 228–249. https://doi.org/10.1016/j.biomaterials.2018.10.030.

Shin, J. I., Ha, J. H., Kim, K. M., Choi, J. G., Park, S. R., Park, H. E. et al. (2023). A novel repeat sequence-based PCR (rep-PCR) using specific repeat sequences of Mycobacterium intracellulare as a DNA fingerprinting. Frontiers in Microbiology 14: 1161194. https://doi.org/10.3389/fmicb.2023.1161194.

Smith, I. (2003). Mycobacterium tuberculosis pathogenesis and molecular determinants of virulence. Clinical Microbiology Reviews 16(3): 463–496. https://doi.org/10.1128/CMR.16.3.463-496.2003.

Souza, W.d (2012). Prokaryotic cells: structural organisation of the cytoskeleton and organelles. Memorias do Instituto Oswaldo Cruz 107(3): 283–293. https://doi.org/10.1590/s0074-02762012000300001.

Spigaglia, P. and Mastrantonio, P. (2003). Evaluation of repetitive element sequence-based PCR as a molecular typing method for Clostridium difficile. Journal of Clinical Microbiology 41(6): 2454–2457. https://doi.org/10.1128/JCM.41.6.2454-2457.2003.

Stewart, E. J. (2012). Growing unculturable bacteria. Journal of Bacteriology 194(16): 4151–4160. https://doi.org/10.1128/JB.00345-12.

Šudomová, M., Berchová-Bímová, K., Mazurakova, A., Šamec, D., Kubatka, P. and Hassan, S. T. S. (2022). Flavonoids target human herpesviruses that infect the nervous system: mechanisms of action and therapeutic insights. Viruses 14(3): 592. https://doi.org/10.3390/v14030592.

Takiishi, T., Fenero, C. I. M. and Câmara, N. O. S. (2017). Intestinal barrier and gut microbiota: Shaping our immune responses throughout life. Tissue Barriers 5(4): e1373208. https://doi.org/10.1080/21688370.2017.1373208.

Thirumdas, R., Kothakota, A., Pandiselvam, R., Bahrami, A. and Barba, F. J. (2021). Role of food nutrients and supplementation in fighting against viral infections and boosting immunity: A review. Trends in Food Science & Technology 110: 66–77. https://doi. org/10.1016/j.tifs.2021.01.069.

Vartoukian, S. R., Adamowska, A., Lawlor, M., Moazzez, R., Dewhirst, F. E. and Wade, W. G. (2016). *In Vitro* cultivation of 'Unculturable' oral bacteria, facilitated by community culture and nedia supplementation with siderophores. PLoS ONE 11(1): e0146926. https://doi.org/10.1371/journal.pone.0146926.

Wall, S. (2019). Prevention of antibiotic resistance - an epidemiological scoping review to identify research categories and knowledge gaps. Global Health Action 12(1): 1756191. https://doi.org/10.1080/16549716.2020.1756191.

Wan, Z., Zheng, J., Zhu, Z., Sang, L., Zhu, J., Luo, S. et al. (2022). Intermediate role of gut microbiota in vitamin B nutrition and its influences on human health. Frontiers in Nutrition 9: 1031502. https://doi.org/10.3389/fnut.2022.1031502.

Wi, Y. M. and Patel, R. (2018). Understanding biofilms and novel approaches to the diagnosis, prevention, and treatment of medical device-associated infections. Infectious Disease Clinics of North America 32(4): 915–929. https://doi.org/10.1016/j.idc.2018.06.009.

Yang, S., Rothman, R. E., Hardick, J., Kuroki, M., Hardick, A., Doshi, V. et al. (2008). Rapid polymerase chain reaction-based screening assay for bacterial biothreat agents. Academic Emergency Medicine: Official Journal of the Society for Academic Emergency Medicine 15(4): 388–392. https://doi.org/10.1111/j.1553-2712.2008.00061.x.

Zhang, Y. J., Li, S., Gan, R. Y., Zhou, T., Xu, D. P. and Li, H. B. (2015). Impacts of gut bacteria on human health and diseases. International Journal of Molecular Sciences 16(4): 7493–7519. https://doi.org/10.3390/ijms16047493.

Zhu, M. Z., Li, N., Zhou, F., Ouyang, J., Lu, D. M., Xu, W. et al. (2020). Microbial bioconversion of the chemical components in dark tea. Food Chemistry 312: 126043. https://doi. org/10.1016/j.foodchem.2019.126043.

Chapter 2
Microbial Diversity and Ecosystems

Introduction

Microbial diversity is "the branch of biology under the faculty of science that deals with the study of microorganisms, i.e., known and unknown microscopic life without true nucleus and those that are present anywhere in the universe".

Microorganisms are the most diverse group of microscopic organisms on Earth, encompassing all bacteria, archaea, fungi, algae, and protozoa. They are found in every environment, from the deepest oceans to the highest mountaintops, and from inside our bodies to the surfaces of objects we touch every day. Microbial diversity is important for a variety of reasons. Microorganisms play essential roles in many ecosystems, such as breaking down organic matter, recycling nutrients, and fixing nitrogen (The Microbial World, 1997; Oren, 2004; Flint, 2020). They also play important roles in human health, both as beneficial symbionts and as pathogens. The study of microbial diversity is a relatively new field, but it has advanced rapidly in recent years due to the development of new technologies, such as next-generation sequencing (Cho and Blaser, 2012; Ursell et al., 2012; Martín et al., 2014; Shreiner et al., 2015; Eloe-Fadrosh and Rasko, 2013; Lozupone et al., 2012; Bertola et al., 2021; Kumar et al., 2022; Aggarwal et al., 2023). This has allowed scientists to identify and characterize a vast array of new microorganisms and to begin to understand the complex relationships between different microbial communities. Microbial diversity can be measured at a variety of levels, from the genetic diversity of individual species to the diversity of microbial communities within a given environment (Institute of Medicine (US) Forum on Microbial Threats, 2012; Gibbons and Gilbert, 2015; Agrawal et al., 2015; Kodera et al., 2022). One common way to measure microbial diversity is to use the Shannon index, which takes into account both the number of different species present and their relative abundance (Woese et al., 1990). The significance of microbial diversity cannot be overstated, as it holds immense importance for a multitude of reasons. Ensuring the resilience of ecosystems is of paramount importance. In the realm of abundant microbial diversity, the intricate interplay between various species manifests in a remarkable phenomenon: the ability of different organisms to harmoniously compensate for one another's shortcomings in the face of disruption. In the realm of industrial processes, it serves as a fundamental cornerstone for a multitude of applications, including but not limited to wastewater treatment and biofuel production. Furthermore, the significance of microbial diversity in relation to human health has a hidden world of where microorganisms thrive, silently orchestrating a symphony of vital functions. These microscopic beings, so often overlooked, hold the key to our digestion and immune fortitude (Abeles et al., 2016; Philippot et al., 2021).

Scope of microbial diversity

The realm of life on Earth known as microbial diversity, which includes bacteria, fungi, protists, and archaea, is enormous and constantly growing (Pace, 1997; Hunter-Cevera et al., 2005; Microbiology in the 21st Century: Where Are We and Where Are We Going?, 2004; Lemke and DeSalle, 2023). There are an estimated tens of quintillions of these microscopic organisms in the world, and they live in almost every kind of environment, from the highest mountain peaks to the deepest ocean trenches (Hemmersbach et al., 2014; Liu et al., 2022). Their profound impact on our planet and our lives is the only thing that can compare to their sheer ubiquity and diversity. Almost every aspect of the biosphere depends on microbiological life, including the cycling of nutrients, nitrogen fixation, atmospheric maintenance, and the biodegradation of pollutants (Horneck et al., 2010; Wang et al., 2017; Shen et al., 2022; Li et al., 2023). In addition, the human body harbours a multifaceted and heterogeneous microbiome that profoundly impacts our overall health and welfare (Ogunrinola et al., 2020; Hou et al., 2022; Clemente et al., 2012; Gilbert et al., 2018; Aggarwal et al., 2023). The potential uses of microbes in industries like biotechnology, agriculture, medicine, and energy production appear limitless as we continue to explore this unexplored region (National Research Council (US) Committee on a National Strategy for Biotechnology in Agriculture, 1987; Microbial Energy Conversion, 2006; Reid and Green, 2012; Vaou et al., 2021). We have come a long way in our understanding of microbial diversity, but we still have a long way to go. The future is brimming with opportunity for advances in a variety of fields, opening the door to a sustainable and prosperous future as new discoveries are made on a regular basis (Gibbons and Gilbert, 2015; Natarajan and Bhatt, 2020; Gonzalez and Aranda, 2023).

The measurement of microbial diversity

Microbial diversity can be measured using a variety of approaches and strategies. Using high-throughput sequencing technologies, such as Next-Generation Sequencing (NGS), to analyze the various microbial species present in a sample is one popular method (Satam et al., 2023; Chaudhari et al., 2023; Nafea et al., 2024). This technique provides insights into the variety and composition of the microbiome by assisting in the identification and measurement of the many microbial populations present in a given environment (Aguiar-Pulido et al., 2016; *National Academies of Sciences, Engineering, and Medicine; Division on Earth and Life Studies; Board on Life Sciences; Board on Environmental Studies and Toxicology; Committee on Advancing Understanding of the Implications of Environmental-Chemical Intera*). Moreover, scientists can create microbial diversity metrics like alpha diversity (within-sample diversity) and beta diversity (between-sample diversity) by analyzing sequencing data using bioinformatic methods. These metrics aid in the comparison of the microbial composition between various samples or habitats as well as the comprehension of the richness and evenness of microbial communities (Galloway-Peña and Hanson, 2019; Douglas et al., 2023). Furthermore, to separate and identify particular microbial species present in a sample, researchers can also use culture-based approaches. However, because many bacteria cannot be cultured in laboratory environments, this conventional method may only be able to capture a portion of the overall diversity of microbes (Avershina et al., 2023; Kapinusova et al., 2023; Wan et al., 2023; Garg et al., 2024). Therefore, to measure microbial diversity, one must thoroughly evaluate the microbial communities and their interactions within a given environment using sophisticated sequencing technologies, bioinformatic analysis, and maybe culture-based methods (Scow et al., 2001).

Colony Forming Units (CFU) count

This is the most basic method for assessing microbial diversity. It involves growing microbes on a solid media plate and counting the number of colonies that form. Each colony represents a single viable cell that has divided and grown into a visible mass (Pires and Bettencourt, 2023; Bhuyan et al., 2023; Sahu et al., 2023; Thomson et al., 2023; Stålfelt et al., 2023; Novakowski et al., 2023; Jumutc et al., 2023; Meyer et al., 2023; Dexter et al., 2024). While simple and relatively inexpensive, CFU only detects microbes that can grow under the specific conditions provided by the media. This means a large portion of the microbial diversity in a sample may be missed (Sieuwerts et al., 2008). Bacterial Colony Forming Units (CFUs) on agar plates are difficult and prone to error to count by hand. Using a novel segmentation method, Brugger et al. (2012) created a colony counting system to distinguish bacterial colonies from blood and other agar plates. This method involved designing a hardware colony counter and writing a novel segmentation algorithm in MATLAB. To sum up, the segmentation stage involved extracting the colony pictures from the blood agar and separating the individual colonies after pre-processing with Top-Hat-filtering to create a homogeneous backdrop. The final count of bacterial colonies was then determined using a Bayes classifier because some colonies could still concatenate to create larger groupings. Automated colony counting of several agar plates with known CFU values of *S. pneumoniae, P. aeruginosa,* and *M. catarrhalis* has demonstrated good performance in evaluating the accuracy and performance of the colony counter (Brugger et al., 2012).

Most Probable Number (MPN) technique

This method is used to estimate the number of viable but uncultivable bacteria in a liquid sample. It involves making a series of dilutions of the sample and inoculating each dilution into a separate growth medium. The number of positive tubes (tubes showing growth) in each dilution is then used to calculate the most probable number of viable bacteria in the original sample. Similar to CFU, MPN also misses a significant portion of the microbial diversity (Ahmad et al., 2017; Moussa et al., 2020).

Biolog EcoPlates

These are commercially available plates that contain a variety of different carbon sources, allowing researchers to assess the metabolic diversity of a microbial community. Each well in the plate contains a different substrate, and the growth of microbes in each well indicates their ability to utilize that particular substrate. By analyzing the pattern of growth across all the wells, researchers can gain insights into the functional diversity of the microbial community (Chojniak et al., 2015; Checcucci et al., 2021).

Community Characterization by Replica Plating (CCRPT)

This method involves replica plating a sample onto a series of different media plates, each designed to select for a specific type of microbe. By comparing the growth patterns on different plates, researchers can get a sense of the diversity of microbes present in the sample (Corlett et al., 1965; Osterblad et al., 1995).

Culture-based methods are still valuable tools for assessing microbial diversity, but they have limitations. They only detect microbes that can be grown in the laboratory, and they may not provide a complete picture of the microbial community present in a sample. However, they can be used to isolate and identify specific microbes of interest, and they can be used in conjunction with culture-independent methods to provide a more comprehensive understanding of microbial diversity (Stoddart, 2011; Vembadi et al., 2019).

The importance of microbial diversity to human health and well-being

Microbes are the most common type of biodiversity resource on Earth. The microbiome of the gut, which is found in the intestines, is home to billions of microbes, the majority of which are found in the human body. These microorganisms are not merely passive inhabitants; rather, they are the building blocks of a complex ecosystem. They interact with one another and with the human host in subtle ways that have an effect on our health throughout our entire lives (Ursell et al., 2012; Bull and Plummer, 2014). Developing microbial resources is important for both using and exploiting genetic resources and for protecting and keeping them safe. This includes finding new organisms and genes that are useful for biotechnology, studying diversity patterns that can be used to track and predict changes in the environment, and understanding the role that the microbiome plays in human health. Microbial diversity study is one of the most important things that is pushing the field of life science forward in the 21st century (CD Genomics, 2024).

A balanced and diverse microbiota is essential for numerous physiological functions. Microbes are responsible for the digestion of complex carbohydrates, fibres, and other nutrients that our systems are unable to process on their own. As a result, they supply us with critical vitamins, minerals, and short-chain fatty acids that nourish our gut cells and assist in maintaining overall immunity and health (Rowland et al., 2018). The gut microbiota plays a crucial role in training and educating the immune system, helping it distinguish between harmful and beneficial substances. A diverse microbiota helps maintain immune tolerance, preventing the immune system from overreacting and causing inflammation (Belkaid and Hand, 2014). The gut microbiota acts as a barrier against harmful pathogens, such as bacteria and viruses, competing for resources and preventing them from colonizing the gut (Pickard et al., 2017). The gut microbiota influences how we extract energy from food and store it. Changes in the composition of the microbiota have been linked to obesity, diabetes, and metabolic disorders (Davis, 2016; Liu et al., 2021). Emerging research suggests a link between the gut microbiota and brain health. The gut microbiome may influence brain development, function, and behaviour through the gut-brain axis, a bidirectional communication pathway between the gut and the central nervous system (Davis, 2016; Appleton, 2018).

Disruptions in the microbiome and potential consequences

Factors like overuse of antibiotics, unhealthy diet, chronic stress, and environmental toxins can disrupt the delicate balance of the gut microbiota, leading to a condition known as dysbiosis. When harmful bacteria overgrow beneficial ones, it can trigger a cascade of negative health consequences (Francino, 2016; Ramirez et al., 2020), including:

1. *Digestive disorders*: Diarrhoea, constipation, Inflammatory Bowel Disease (IBD), and Irritable Bowel Syndrome (IBS) have all been linked to dysbiosis (Menees and Chey, 2018; Shaikh et al., 2023).

2. *Autoimmune diseases*: A growing body of research suggests a link between dysbiosis and autoimmune diseases like rheumatoid arthritis, type 1 diabetes, and multiple sclerosis (Mangalam et al., 2021; Mousa et al., 2022).

3. *Neurological disorders*: Emerging evidence suggests that alterations in the gut microbiota may contribute to the development of neurological disorders like Autism Spectrum Disorder (ASD), Parkinson's disease, and Alzheimer's disease (Suganya and Koo, 2020; Ullah et al., 2023).

The promise of microbiome medicine

Given the profound impact of microbial diversity on human health, harnessing the potential of the microbiome for therapeutic purposes is a rapidly evolving field known as microbiome medicine (Dietert, 2021; Zhang et al., 2022; Gebrayel et al., 2022). Researchers are exploring various strategies to manipulate the gut microbiota for therapeutic benefit, including:

a) *Faecal Microbiota Transplant (FMT)*: In this procedure, healthy donor stool is introduced into the gut of a recipient with dysbiosis, aiming to restore a healthy microbial balance (Gupta et al., 2016; Kim and Gluck, 2019).

b) *Prebiotics and probiotics*: Prebiotics are non-digestible fibres that serve as food for beneficial gut microbes, while probiotics are live microorganisms that provide direct health benefits. Both are being explored for their potential to promote a healthy gut microbiota (Davani-Davari et al., 2019; Wu and Chiou, 2021; Roy and Dhaneshwar, 2023).

c) *Next-generation therapies*: Researchers are developing targeted therapies based on our growing understanding of the microbiome, such as personalized probiotics and bacteriophages (viruses that specifically target and kill harmful bacteria) (Sulakvelidze et al., 2001; Lin et al., 2017; Yadav and Chauhan, 2021; Gulliver et al., 2022).

Strategies for conserving microbial diversity

Microbial diversity, often hidden from the naked eye, plays a crucial role in maintaining healthy ecosystems and providing essential services for life on Earth. Conserving this remarkable diversity is paramount, and researchers are developing a range of strategies to achieve this (Colwell, 1997; Delgado-Baquerizo, 2016; Eke et al., 2023; Manyi-Loh and Lues, 2023). Here are some key approaches:

In-situ conservation

- *Habitat protection*: This involves setting aside areas like wetlands, coral reefs, and forests where unique microbial communities thrive. Stricter regulations on deforestation, unsustainable agricultural practices, and pollution control are crucial (Hails, 1997).

- *Ecosystem restoration*: Restoring degraded environments can foster the return of lost microbial diversity. This includes replanting native vegetation, improving soil health, and reducing contamination. In addition, mixing of biopolymers such as cellulose promotes growth of slow-growing bacteria and fungi (Docherty and Gutknecht, 2019).

- *Sustainable management*: Implementing sustainable practices in agriculture, forestry, and other sectors like responsible waste management and responsible use of antibiotics helps maintain healthy microbial communities (Visca et al., 2024) in working landscapes.

Ex-situ conservation

- *Culture collections*: Specialized facilities store diverse microbial strains in cryopreserved or lyophilized states, preserving them for future use or research.

- *Metagenomic repositories*: Databases store genetic information from entire microbial communities, enabling researchers to analyze and potentially recreate them in future restoration efforts (Refer: ATCC USA; DSMZ Germany; CSIR-IMTECH India; DBT-NCCS India).

Other strategies

- *Microbial prospecting*: Identifying and characterizing novel microbes with unique and potentially beneficial properties for medicine, agriculture, or bioremediation can incentivize their conservation (Strobel and Daisy, 2003).

- *Public awareness*: Raising public awareness about the importance of microbial diversity and promoting responsible practices that protect their habitats is crucial for long-term sustainability (Figure 1) (Rappuoli et al., 2023).

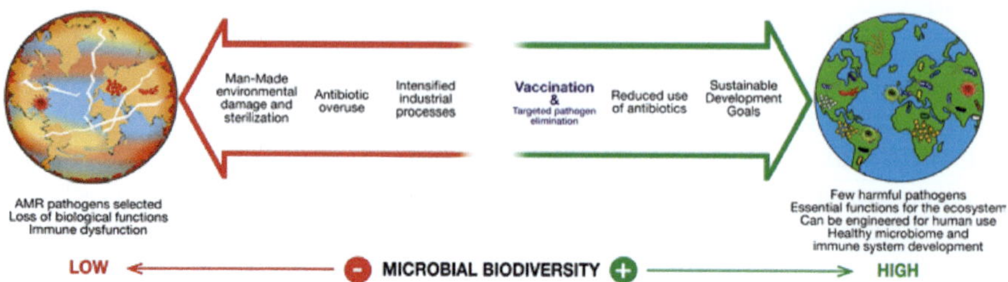

Figure 1. Barriers for conservation of Microbial diversity (Adopted from Rappuoli et al., 2023).

References

Abeles, S. R., Jones, M. B., Santiago-Rodriguez, T. M., Ly, M., Klitgord, N., Yooseph, S. et al. (2016). Microbial diversity in individuals and their household contacts following typical antibiotic courses. Microbiome 4(1): 39. https://doi.org/10.1186/s40168-016-0187-9.

Aggarwal, N., Kitano, S., Puah, G. R. Y., Kittelmann, S., Hwang, I. Y. and Chang, M. W. (2023). Microbiome and human health: current understanding, engineering, and enabling technologies. Chemical Reviews 123(1): 31–72. https://doi.org/10.1021/acs.chemrev.2c00431.

Agrawal, P. K., Agrawal, S. and Shrivastava, R. (2015). Modern molecular approaches for analyzing microbial diversity from mushroom compost ecosystem. 3 Biotech 5(6): 853–866. https://doi.org/10.1007/s13205-015-0289-2.

Aguiar-Pulido, V., Huang, W., Suarez-Ulloa, V., Cickovski, T., Mathee, K. and Narasimhan, G. (2016). Metagenomics, metatranscriptomics, and metabolomics approaches for microbiome analysis. Evolutionary Bioinformatics Online 12(Suppl 1): 5–16. https://doi.org/10.4137/EBO.S36436.

Ahmad, F., Stedtfeld, R. D., Waseem, H., Williams, M. R., Cupples, A. M., Tiedje, J. M. et al. (2017). Most probable number - loop mediated isothermal amplification (MPN-LAMP) for quantifying waterborne pathogens in < 25 min. J. Microbiol. Meth. 132: 27–33. https://doi.org/10.1016/j.mimet.2016.11.010.

Appleton, J. (2018). The gut-brain axis: influence of microbiota on mood and mental health. Integrative Medicine (Encinitas, Calif.) 17(4): 28–32.

Avershina, E., Khezri, A. and Ahmad, R. (2023). Clinical diagnostics of bacterial infections and their resistance to antibiotics-current state and whole genome sequencing implementation perspectives. Antibiotics (Basel, Switzerland) 12(4): 781. https://doi.org/10.3390/antibiotics12040781.

Belkaid, Y. and Hand, T. W. (2014). Role of the microbiota in immunity and inflammation. Cell 157(1): 121–141. https://doi.org/10.1016/j.cell.2014.03.011.

Bertola, M., Ferrarini, A. and Visioli, G. (2021). Improvement of soil microbial diversity through sustainable agricultural practices and its evaluation by -omics approaches: a perspective for the environment, food quality and human safety. Microorganisms 9(7): 1400. https://doi.org/10.3390/microorganisms9071400.

Bhuyan, S., Yadav, M., Giri, S. J., Begum, S., Das, S., Phukan, A. et al. (2023). Microliter spotting and micro-colony observation: A rapid and simple approach for counting bacterial colony forming units. Journal of Microbiological Methods 207: 106707. https://doi.org/10.1016/j.mimet.2023.106707.

Brugger, S. D., Baumberger, C., Jost, M., Jenni, W., Brugger, U. and Mühlemann, K. (2012). Automated counting of bacterial colony forming units on agar plates. PloS One 7(3): e33695. https://doi.org/10.1371/journal.pone.0033695.

Bull, M. J. and Plummer, N. T. (2014). Part 1: The human gut microbiome in health and disease. Integrative Medicine (Encinitas, Calif.) 13(6): 17–22.

CD Genomics. Significance of Microbial Diversity Research. https://www.cd-genomics.com/microbioseq/microbial-diversity-significance-and-research-methodology.html Assessed on 20-02-2024.

Chaudhari, H. G., Prajapati, S., Wardah, Z. H., Raol, G., Prajapati, V., Patel, R. et al. (2023). Decoding the microbial universe with metagenomics: a brief insight. Frontiers in Genetics 14: 1119740. https://doi.org/10.3389/fgene.2023.1119740.

Checcucci, A., Luise, D., Modesto, M., Correa, F., Bosi, P., Mattarelli, P. et al. (2021). Assessment of Biolog EcoplateTM method for functional metabolic diversity of aerotolerant pig fecal microbiota. Applied Microbiology and Biotechnology 105(14-15): 6033–6045. https://doi.org/10.1007/s00253-021-11449-x.

Cho, I. and Blaser, M. J. (2012). The human microbiome: at the interface of health and disease. Nature Reviews. Genetics 13(4): 260–270. https://doi.org/10.1038/nrg3182.

Chojniak, J., Jałowiecki, Ł., Dorgeloh, E., Hegedusova, B., Ejhed, H., Magnér, J. et al. (2015). Application of the BIOLOG system for characterization of Serratia marcescens ss marcescens isolated from onsite wastewater technology (OSWT). Acta Biochimica Polonica 62(4): 799–805. https://doi.org/10.18388/abp.2015_1138.

Clemente, J. C., Ursell, L. K., Parfrey, L. W. and Knight, R. (2012). The impact of the gut microbiota on human health: an integrative view. Cell 148(6): 1258–1270. https://doi.org/10.1016/j.cell.2012.01.035.

Colwell, R. R. (1997). Microbial diversity: the importance of exploration and conservation. Journal of Industrial Microbiology & Biotechnology 18(5): 302–307. https://doi.org/10.1038/sj.jim.2900390.

Corlett, D. A., Jr, Lee, J. S. and Sinnhuber, R. O. (1965). Application of replica plating and computer analysis for rapid identification of bacteria in some foods. I. Identification scheme. Applied Microbiology 13(5): 808–817. https://doi.org/10.1128/am.13.5.808-817.1965.

Davani-Davari, D., Negahdaripour, M., Karimzadeh, I., Seifan, M., Mohkam, M., Masoumi, S. J. et al. (2019). Prebiotics: Definition, types, sources, mechanisms, and clinical applications. Foods (Basel, Switzerland) 8(3): 92. https://doi.org/10.3390/foods8030092.

Davis, C. D. (2016). The gut microbiome and its role in obesity. Nutrition Today 51(4): 167–174. https://doi.org/10.1097/NT.0000000000000167.

Dekaboruah, E., Suryavanshi, M. V., Chettri, D. and Verma, A. K. (2020). Human microbiome: an academic update on human body site specific surveillance and its possible role. Archives of Microbiology 202(8): 2147–2167. https://doi.org/10.1007/s00203-020-01931-x.

Delgado-Baquerizo, M., Maestre, F. T., Reich, P. B., Jeffries, T. C., Gaitan, J. J., Encinar, D. et al. (2016). Microbial diversity drives multifunctionality in terrestrial ecosystems. Nature Communications 7: 10541. https://doi.org/10.1038/ncomms10541.

Demain, A. L. and Adrio, J. L. (2008). Contributions of microorganisms to industrial biology. Molecular Biotechnology 38(1): 41–55. https://doi.org/10.1007/s12033-007-0035-z.

Dexter, F., Walker, K. M., Brindeiro, C. T., Loftus, C. P., Banguid, C. C. L. and Loftus, R. W. (2024). A threshold of 100 or more colony-forming units on the anesthesia machine predicts bacterial pathogen detection: a retrospective laboratory-based analysis. Un seuil de 100 unités de formation de colonie ou plus sur l'appareil d'anesthésie prédit la détection d'agents pathogènes bactériens : une analyse rétrospective en laboratoire. Canadian Journal of Anaesthesia = Journal canadien d'anesthesie, 10.1007/s12630-024-02707-3. Advance online publication. https://doi.org/10.1007/s12630-024-02707-3.

Dietert, R. R. (2021). Microbiome first medicine in health and safety. Biomedicines 9(9): 1099. https://doi.org/10.3390/biomedicines9091099.

Docherty, K. M. and Gutknecht, J. L. M. (2019). Soil microbial restoration strategies for promoting climate-ready prairie ecosystems. Ecological Applications: A Publication of the Ecological Society of America 29(3): e01858. https://doi.org/10.1002/eap.1858.

Douglas, G. M., Kim, S., Langille, M. G. I. and Shapiro, B. J. (2023). Efficient computation of contributional diversity metrics from microbiome data with FuncDiv. Bioinformatics (Oxford, England) 39(1): btac809. https://doi.org/10.1093/bioinformatics/btac809.

Eke, M., Tougeron, K., Hamidovic, A., Tinkeu, L. S. N., Hance, T. and Renoz, F. (2023). Deciphering the functional diversity of the gut microbiota of the black soldier fly (Hermetia illucens): recent advances and future challenges. Animal Microbiome 5(1): 40. https://doi.org/10.1186/s42523-023-00261-9.

Eloe-Fadrosh, E. A. and Rasko, D. A. (2013). The human microbiome: from symbiosis to pathogenesis. Annual Review of Medicine 64: 145–163. https://doi.org/10.1146/annurev-med-010312-133513.

Flint, H. J. (2020). Micro-organisms and the microbiome. Why Gut Microbes Matter: Understanding Our Microbiome, 1–8. https://doi.org/10.1007/978-3-030-43246-1_1.

Francino, M. P. (2016). Antibiotics and the human gut microbiome: dysbioses and accumulation of resistances. Frontiers in Microbiology 6: 1543. https://doi.org/10.3389/fmicb.2015.01543.

Galloway-Peña, J. and Hanson, B. (2020). Tools for analysis of the microbiome. Digestive Diseases and Sciences 65(3): 674–685. https://doi.org/10.1007/s10620-020-06091-y.

Garg, D., Patel, N., Rawat, A. and Rosado, A. S. (2024). Cutting edge tools in the field of soil microbiology. Current Research in Microbial Sciences 6: 100226. https://doi.org/10.1016/j.crmicr.2024.100226.

Gebrayel, P., Nicco, C., Al Khodor, S., Bilinski, J., Caselli, E., Comelli, E. M. et al. (2022). Microbiota medicine: towards clinical revolution. Journal of Translational Medicine 20(1): 111. https://doi.org/10.1186/s12967-022-03296-9.

Gibbons, S. M. and Gilbert, J. A. (2015). Microbial diversity—exploration of natural ecosystems and microbiomes. Current Opinion in Genetics & Development 35: 66–72. https://doi.org/10.1016/j.gde.2015.10.003.

Gilbert, J. A., Blaser, M. J., Caporaso, J. G., Jansson, J. K., Lynch, S. V. and Knight, R. (2018). Current understanding of the human microbiome. Nature Medicine 24(4): 392–400. https://doi.org/10.1038/nm.4517.

Gonzalez, J. M. and Aranda, B. (2023). Microbial growth under limiting conditions-future perspectives. Microorganisms 11(7): 1641. https://doi.org/10.3390/microorganisms11071641.

Gulliver, E. L., Young, R. B., Chonwerawong, M., D'Adamo, G. L., Thomason, T., Widdop, J. T. et al. (2022). Review article: the future of microbiome-based therapeutics. Alimentary Pharmacology & Therapeutics 56(2): 192–208. https://doi.org/10.1111/apt.17049.

Gupta, S., Allen-Vercoe, E. and Petrof, E. O. (2016). Fecal microbiota transplantation: in perspective. Therapeutic Advances in Gastroenterology 9(2): 229–239. https://doi.org/10.1177/1756283X15607414.

Hails, A. J. (1997). Wetlands, Biodiversity and the Ramsar Convention. Ramsar Convention Bureau Ministry of Environment and Forest, India. The Ramsar Convention Bureau, Gland, Switzerland.

Hemmersbach, R., Simon, A., Waßer, K., Hauslage, J., Christianen, P. C., Albers, P. W. et al. (2014). Impact of a high magnetic field on the orientation of gravitactic unicellular organisms—a critical consideration about the application of magnetic fields to mimic functional weightlessness. Astrobiology 14(3): 205–215. https://doi.org/10.1089/ast.2013.1085.

Horneck, G., Klaus, D. M. and Mancinelli, R. L. (2010). Space microbiology. Microbiology and Molecular Biology Reviews: MMBR 74(1): 121–156. https://doi.org/10.1128/MMBR.00016-09.

Hou, K., Wu, Z. X., Chen, X. Y., Wang, J. Q., Zhang, D., Xiao, C. et al. (2022). Microbiota in health and diseases. Signal Transduction and Targeted Therapy 7(1): 135. https://doi.org/10.1038/s41392-022-00974-4.

Hunter-Cevera, J., Karl, D. and Buckley, M. (2005). Marine Microbial Diversity: The Key to Earth's Habitability: This report is based on a colloquium, sponsored by the American Academy of Microbiology, held April 8–10, 2005, in San Francisco, California. Washington (DC): American Society for Microbiology; 2005. Available from: https://www.ncbi.nlm.nih.gov/books/NBK559439/ doi: 10.1128/AAMCol.8Apr.2005.

Institute of Medicine (US) Forum on Microbial Threats. The Social Biology of Microbial Communities: Workshop Summary. Washington (DC): National Academies Press (US); 2012. Workshop Overview. Available from: https://www.ncbi.nlm.nih.gov/books/NBK154539/.

Jumutc, V., Suponenkovs, A., Bondarenko, A., Bļizņuks, D. and Lihachev, A. (2023). Hybrid approach to colony-forming unit counting problem using multi-loss U-Net reformulation. Sensors (Basel, Switzerland) 23(19): 8337. https://doi.org/10.3390/s23198337.

Kapinusova, G., Lopez Marin, M. A. and Uhlik, O. (2023). Reaching unreachables: Obstacles and successes of microbial cultivation and their reasons. Frontiers in Microbiology 14: 1089630. https://doi.org/10.3389/fmicb.2023.1089630.

Kim, K. O. and Gluck, M. (2019). Fecal microbiota transplantation: an update on clinical practice. Clinical endoscopy, 52(2), 137–143. https://doi.org/10.5946/ce.2019.009.

Kodera, S. M., Das, P., Gilbert, J. A. and Lutz, H. L. (2022). Conceptual strategies for characterizing interactions in microbial communities. iScience 25(2): 103775. https://doi.org/10.1016/j.isci.2022.103775.

Kumar, R., Sood, U., Kaur, J., Anand, S., Gupta, V., Patil, K. S. et al. (2022). The rising dominance of microbiology: what to expect in the next 15 years?. Microbial Biotechnology 15(1): 110–128. https://doi.org/10.1111/1751-7915.13953.

Lemke, M. and DeSalle, R. (2023). The next generation of microbial ecology and its importance in environmental sustainability. Microbial Ecology 85(3): 781–795. https://doi.org/10.1007/s00248-023-02185-y.

Li, Y., Xiong, L., Yu, H., Zeng, K., Wei, Y., Li, H. et al. (2023). Function and distribution of nitrogen-cycling microbial communities in the Napahai plateau wetland. Archives of Microbiology 205(11): 357. https://doi.org/10.1007/s00203-023-03695-6.

Lin, D. M., Koskella, B. and Lin, H. C. (2017). Phage therapy: An alternative to antibiotics in the age of multi-drug resistance. World Journal of Gastrointestinal Pharmacology and Therapeutics 8(3): 162–173. https://doi.org/10.4292/wjgpt.v8.i3.162.

Liu, B. N., Liu, X. T., Liang, Z. H. and Wang, J. H. (2021). Gut microbiota in obesity. World Journal of Gastroenterology 27(25): 3837–3850. https://doi.org/10.3748/wjg.v27.i25.3837.

Liu, Y., Zhang, Z., Ji, M., Hu, A., Wang, J., Jing, H. et al. (2022). Comparison of prokaryotes between Mount Everest and the Mariana Trench. Microbiome 10(1): 215. https://doi.org/10.1186/s40168-022-01403-y.

Lozupone, C. A., Stombaugh, J. I., Gordon, J. I., Jansson, J. K. and Knight, R. (2012). Diversity, stability and resilience of the human gut microbiota. Nature 489(7415): 220–230. https://doi.org/10.1038/nature11550.

Mangalam, A. K., Yadav, M. and Yadav, R. (2021). The emerging world of microbiome in autoimmune disorders: opportunities and challenges. Indian Journal of Rheumatology 16(1): 57–72. https://doi.org/10.4103/injr.injr_210_20.

Manyi-Loh, C. E. and Lues, R. (2023). A South African perspective on the microbiological and chemical quality of meat: plausible public health implications. Microorganisms 11(10): 2484. https://doi.org/10.3390/microorganisms11102484.

Martín, R., Miquel, S., Langella, P. and Bermúdez-Humarán, L. G. (2014). The role of metagenomics in understanding the human microbiome in health and disease. Virulence 5(3): 413–423. https://doi.org/10.4161/viru.27864.

Menees, S. and Chey, W. (2018). The gut microbiome and irritable bowel syndrome. F1000Research 7: F1000 Faculty Rev-1029. https://doi.org/10.12688/f1000research.14592.1.

Meyer, C. T., Lynch, G. K., Stamo, D. F., Miller, E. J., Chatterjee, A. and Kralj, J. M. (2023). A high-throughput and low-waste viability assay for microbes. Nature Microbiology 8(12): 2304–2314. https://doi.org/10.1038/s41564-023-01513-9.

Microbial Energy Conversion: This report is based on a colloquium, sponsored by the American Academy of Microbiology, convened March 10–12, 2006, in San Francisco, California. Washington (DC): American Society for Microbiology; 2006. Available from: https://www.ncbi.nlm.nih.gov/books/NBK563531/ doi: 10.1128/AAMCol.10Mar.2006.

Microbiology in the 21st Century: Where Are We and Where Are We Going? This report is based on a colloquium sponsored by the American Academy of Microbiology held September 5–7, 2003, in Charleston, South Carolina. Washington (DC): American Society for Microbiology; 2004. Available from: https://www.ncbi.nlm.nih.gov/books/NBK560448/ doi: 10.1128/AAMCol.5Sept.2003.

Mousa, W. K., Chehadeh, F. and Husband, S. (2022). Microbial dysbiosis in the gut drives systemic autoimmune diseases. Frontiers in Immunology 13: 906258. https://doi.org/10.3389/fimmu.2022.906258.

Moussa, M., Marcelino, I., Richard, V., Guerlotté, J. and Talarmin, A. (2020). An optimized most probable number (MPN) method to assess the number of thermophilic free-living amoebae (FLA) in water samples. Pathogens (Basel, Switzerland) 9(5): 409. https://doi.org/10.3390/pathogens9050409.

Nafea, A. M., Wang, Y., Wang, D., Salama, A. M., Aziz, M. A., Xu, S. et al. (2024). Application of next-generation sequencing to identify different pathogens. Frontiers in Microbiology 14: 1329330. https://doi.org/10.3389/fmicb.2023.1329330.

Natarajan, A. and Bhatt, A. S. (2020). Microbes and microbiomes in 2020 and beyond. Nature Communications 11(1): 4988. https://doi.org/10.1038/s41467-020-18850-6.

National Academies of Sciences, Engineering, and Medicine; Division on Earth and Life Studies; Board on Life Sciences; Board on Environmental Studies and Toxicology; Committee on Advancing Understanding of the Implications of Environmental-Chemical Interactions with the Human Microbiome. Environmental Chemicals, the Human Microbiome, and Health Risk: A Research Strategy. Washington (DC): National Academies Press (US); 2017 Dec 29. 4, Current Methods for Studying the Human Microbiome. Available from: https://www.ncbi.nlm.nih.gov/books/NBK481559/.

National Research Council (US) Committee on a National Strategy for Biotechnology in Agriculture. Agricultural Biotechnology: Strategies for National Competitiveness. Washington (DC): National Academies Press (US); 1987. 2, Scientific Aspects. Available from: https://www.ncbi.nlm.nih.gov/books/NBK217989/.

Novakowski, K. E., Loukov, D. and Bowdish, D. M. E. (2023). Bacterial binding, phagocytosis, and killing capacity: measurements using colony forming units. Methods in Molecular Biology (Clifton, N.J.) 2692: 1–13. https://doi.org/10.1007/978-1-0716-3338-0_1.

Ogunrinola, G. A., Oyewale, J. O., Oshamika, O. O. and Olasehinde, G. I. (2020). The human microbiome and its impacts on health. International Journal of Microbiology 2020: 8045646. https://doi.org/10.1155/2020/8045646.

Oren, A. (2004). Prokaryote diversity and taxonomy: current status and future challenges. Philosophical Transactions of the Royal Society of London. Series B, Biological Sciences 359(1444): 623–638. https://doi.org/10.1098/rstb.2003.1458.

Osterblad, M., Leistevuo, T. and Huovinen, P. (1995). Screening for antimicrobial resistance in fecal samples by the replica plating method. Journal of Clinical Microbiology 33(12): 3146–3149. https://doi.org/10.1128/jcm.33.12.3146-3149.1995.

Pace, N. R. (1997). A molecular view of microbial diversity and the biosphere. Science (New York, N.Y.) 276(5313): 734–740. https://doi.org/10.1126/science.276.5313.734.

Philippot, L., Griffiths, B. S. and Langenheder, S. (2021). Microbial community resilience across ecosystems and multiple disturbances. Microbiology and Molecular Biology Reviews : MMBR 85(2): e00026-20. https://doi.org/10.1128/MMBR.00026-20.

Pickard, J. M., Zeng, M. Y., Caruso, R. and Núñez, G. (2017). Gut microbiota: Role in pathogen colonization, immune responses, and inflammatory disease. Immunological Reviews 279(1): 70–89. https://doi.org/10.1111/imr.12567.

Pires, D. and Bettencourt, P. J. G. (2023). Micro-colony forming unit assay for efficacy evaluation of vaccines against tuberculosis. Journal of Visualized Experiments : JoVE (197): 10.3791/65447. https://doi.org/10.3791/65447.

Ramirez, J., Guarner, F., Bustos Fernandez, L., Maruy, A., Sdepanian, V. L. and Cohen, H. (2020). Antibiotics as major disruptors of gut microbiota. Frontiers in Cellular and Infection Microbiology 10: 572912. https://doi.org/10.3389/fcimb.2020.572912.

Rappuoli, R., Young, P., Ron, E., Pecetta, S. and Pizza, M. (2023). Save the microbes to save the planet. A call to action of the International Union of the Microbiological Societies (IUMS). One Health Outlook 5(1): 5. https://doi.org/10.1186/s42522-023-00077-2.

Reid, A. and Greene, S. E. (2012). How Microbes Can Help Feed the World: Report on an American Academy of Microbiology Colloquium Washington, DC // December 2012. Washington (DC): American Society for Microbiology; 2012. Available from: https://www.ncbi.nlm.nih.gov/books/NBK559436/ doi: 10.1128/AAMCol.Dec.2012.

Rowland, I., Gibson, G., Heinken, A., Scott, K., Swann, J., Thiele, I. et al. (2018). Gut microbiota functions: metabolism of nutrients and other food components. European Journal of Nutrition 57(1): 1–24. https://doi.org/10.1007/s00394-017-1445-8.

Roy, S. and Dhaneshwar, S. (2023). Role of prebiotics, probiotics, and synbiotics in management of inflammatory bowel disease: Current perspectives. World Journal of Gastroenterology 29(14): 2078–2100. https://doi.org/10.3748/wjg.v29.i14.2078.

Sahu, S. R., Utkalaja, B. G., Patel, S. K. and Acharya, N. (2023). Spot assay and colony forming unit (CFU) analyses-based sensitivity test for Candida albicans and Saccharomyces cerevisiae. Bio-protocol 13(21): e4872. https://doi.org/10.21769/BioProtoc.4872.

Satam, H., Joshi, K., Mangrolia, U., Waghoo, S., Zaidi, G., Rawool, S. et al. (2023). Next-generation sequencing technology: current trends and advancements. Biology 12(7): 997. https://doi.org/10.3390/biology12070997.

Scow, K. M., Schwartz, E., Johnson, M. J. and Macalady, J. L. (2001). Microbial biodiversity, measurement of Encyclopedia of Biodiversity, Volume 4, Academic Press. pp. 177–190.

Shaikh, S. D., Sun, N., Canakis, A., Park, W. Y. and Weber, H. C. (2023). Irritable bowel syndrome and the gut microbiome: a comprehensive review. Journal of Clinical Medicine 12(7): 2558. https://doi.org/10.3390/jcm12072558.

Shen, M., Song, B., Zhou, C., Almatrafi, E., Hu, T., Zeng, G. et al. (2022). Recent advances in impacts of microplastics on nitrogen cycling in the environment: A review. The Science of the Total Environment 815: 152740. https://doi.org/10.1016/j.scitotenv.2021.152740.

Shreiner, A. B., Kao, J. Y. and Young, V. B. (2015). The gut microbiome in health and in disease. Current Opinion in Gastroenterology 31(1): 69–75. https://doi.org/10.1097/MOG.0000000000000139.

Sieuwerts, S., de Bok, F. A., Mols, E., de vos, W. M. and Vlieg, J. E. (2008). A simple and fast method for determining colony forming units. Lett. Appl. Microbiol. 47(4): 275–8. doi: 10.1111/j.1472-765X.2008.02417.x.

Stålfelt, F., Svensson Malchau, K., Björn, C., Mohaddes, M. and Erichsen Andersson, A. (2023). Can particle counting replace conventional surveillance for airborne bacterial contamination assessments? A systematic review using narrative synthesis. American Journal of Infection Control 51(12): 1417–1424. https://doi.org/10.1016/j.ajic.2023.05.004.

Stoddart, M. J. (2011). Cell viability assays: introduction. pp. 1–6. *In*: Stoddart, M. J. (ed.). Mammalian Cell Viability: Methods and Protocols. Totowa, NJ: Humana Press (Methods in Molecular Biology). doi.org/10.1007/978-1-61779-108-6_1.

Strobel, G. and Daisy, B. (2003). Bioprospecting for microbial endophytes and their natural products. Microbiology and Molecular Biology Reviews: MMBR 67(4): 491–502. https://doi.org/10.1128/MMBR.67.4.491-502.2003.

Suganya, K. and Koo, B. S. (2020). Gut-brain axis: role of gut microbiota on neurological disorders and how probiotics/prebiotics beneficially modulate microbial and immune pathways to improve brain functions. International Journal of Molecular Sciences 21(20): 7551. https://doi.org/10.3390/ijms21207551.

Sulakvelidze, A., Alavidze, Z. and Morris, J. G., Jr (2001). Bacteriophage therapy. Antimicrobial Agents and Chemotherapy 45(3): 649–659. https://doi.org/10.1128/AAC.45.3.649-659.2001.

The Microbial World: Foundation of the Biosphere: This report is based on an American Academy of Microbiology colloquium held January 19–21, 1996, in Palm Coast, Florida. The colloquium was supported by the National Science Foundation, the National Oceanic and Atmospheric Administration of the U.S. Department of Commerce, the U.S. Department of Energy, and the American Society for Microbiology. Washington (DC): American Society for Microbiology; 1997. Available from: https://www.ncbi.nlm.nih.gov/books/NBK562919/ doi: 10.1128/AAMCol.19Jan.1996.

Thompson, E. N., Carlino, M. J., Scanlon, V. M., Grimes, H. L. and Krause, D. S. (2023). Assay optimization for the objective quantification of human multilineage colony-forming units. Experimental Hematology 124: 36–44.e3. https://doi.org/10.1016/j.exphem.2023.05.007.

Ullah, H., Arbab, S., Tian, Y., Liu, C. Q., Chen, Y., Qijie, L. et al. (2023). The gut microbiota-brain axis in neurological disorder. Frontiers in Neuroscience 17: 1225875. https://doi.org/10.3389/fnins.2023.1225875.

Ursell, L. K., Metcalf, J. L., Parfrey, L. W. and Knight, R. (2012). Defining the human microbiome. Nutrition Reviews 70 Suppl 1(Suppl 1): S38–S44. https://doi.org/10.1111/j.1753-4887.2012.00493.x.

Vaou, N., Stavropoulou, E., Voidarou, C., Tsigalou, C. and Bezirtzoglou, E. (2021). Towards advances in medicinal plant antimicrobial activity: a review study on challenges and future perspectives. Microorganisms 9(10): 2041. https://doi.org/10.3390/microorganisms9102041.

Vembadi, A., Menachery, A. and Qasaimeh, M.A. (2019) Cell cytometry: review and perspective on biotechnological advances. Front. Bioeng. Biotechnol. 7: 147. https://doi.org/10.3389/fbioe.2019.00147.

Visca, A., Di Gregorio, L., Clagnan, E. and Bevivino, A. (2024). Sustainable strategies: Nature-based solutions to tackle antibiotic resistance gene proliferation and improve agricultural productivity and soil quality. Environmental Research 248: 118395. Advance Online Publication. https://doi.org/10.1016/j.envres.2024.118395.

Wan, X., Yang, Q., Wang, X., Bai, Y. and Liu, Z. (2023). Isolation and cultivation of human gut microorganisms: a review. Microorganisms 11(4): 1080. https://doi.org/10.3390/microorganisms11041080.

Wang, Y., Hatt, J. K., Tsementzi, D., Rodriguez-R, L. M., Ruiz-Pérez, C. A., Weigand, M. R. et al. (2017). Quantifying the importance of the rare biosphere for microbial community response to organic pollutants in a freshwater ecosystem. Applied and Environmental Microbiology 83(8): e03321-16. https://doi.org/10.1128/AEM.03321-16.

Woese, C. R., Kandler, O. and Wheelis, M. L. (1990). Towards a natural system of organisms: proposal for the domains Archaea, Bacteria, and Eukarya. Proceedings of the National Academy of Sciences 87(12): 4576–4579.

Wu, H. and Chiou, J. (2021). Potential benefits of probiotics and prebiotics for coronary heart disease and stroke. Nutrients 13(8): 2878. https://doi.org/10.3390/nu13082878.

Yadav, M. and Chauhan, N. S. (2021). Microbiome therapeutics: exploring the present scenario and challenges. Gastroenterology Report 10: goab046. https://doi.org/10.1093/gastro/goab046.

Zhang, Y., Zhou, L., Xia, J., Dong, C. and Luo, X. (2022). Human microbiome and its medical applications. Frontiers in Molecular Biosciences 8: 703585. https://doi.org/10.3389/fmolb.2021.703585.

Chapter 3
BIOLOG
The Advanced Assessment Methods for the Implementation of Microbe Hunting

Introduction

The vast, unseen world of microbes hums beneath our feet, teeming with diversity and holding immense potential for solving some of humanity's most pressing challenges (The Microbial World, 1996; van der Heijden et al., 2008; Gibbons and Gilbert, 2015; Lane, 2015; Huang, 2023). From unlocking new medicines and biofuels to mitigating climate change and remediating polluted environments, these tiny lifeforms offer untold possibilities (Jeswani et al., 2020). But how do we navigate this hidden biosphere? Enter the thrilling pursuit of "microbe hunting," where biochemical assessment methods serve as our compass and key. This chapter delves into the exciting world of these powerful tool, for example, BIOLOG, unveiling their secrets and highlighting their crucial role in the successful hunt (Németh et al., 2021; Checcucci et al., 2021; Gryta et al., 2014; Pierce et al., 2014; Weber et al., 2007; Sohn et al., 2021; Lim et al., 2024; Tao et al., 2024) for beneficial microbes (Smith, 1984; Lipkin, 2010; Arrigo and Lipkin, 2012; Kapinusova et al., 2023).

Principle of BIOLOG

The Biolog method is a powerful tool in microbial ecology, offering a rapid and standardized approach to assessing the metabolic potential and diversity of microbial communities. Unlike traditional methods that rely on isolating and culturing individual microorganisms, Biolog utilizes a phenotypic fingerprinting technique. This means it analyzes the functional capabilities of a microbial community based on its ability to utilize a wide range of carbon substrates. When discrete test responses are carried out inside a 96-well microplate, pathogenic and environmental microorganisms that generate a "metabolic fingerprint" can be identified using Biolog's phenotypic profiling technique (Source: BIOLOG Website; https://www.biolog.com/).

Statistical analysis

The absorbance value of a control well, which does not include any substrate, is subtracted from the absorbance value of each well that contains a substrate first. This is done before any statistical analyses are performed. This is how one obtains the figure that is referred to as the net absorbance. As a result of the fact that the absorbance value of the control well can be marginally greater than the absorbance values of certain substrates, the absorbance value of the substrate that was utilized the least became the subtraction. Therefore, it is possible to prevent having findings that have negative absorbance values. Numerous methods are available for carrying out statistical analysis of the results (Hitzl et al., 1997; Kelly and Tate, 1998). Before making a decision, it is necessary to determine the number of absorbance measurements that were carried out. As it is necessary to select an ideal incubation duration, which may vary depending on the soil community, a single measurement is not a good answer. This is because of varied soil communities. Whereas the generation of colour has not yet been noticed in certain plates (for the communities that are the least active), the colour development may have reached the point of saturation in other plates. There is also a shift in the structure of the community that occurs throughout the incubation period. This is the reason why a single measurement that is made too late reflects the reduction of tetrazolium dye that is carried out by only a few dominating microbial species of bacteria. By performing a number of measurements throughout the incubation process and then selecting for statistical analysis those readings that exhibit approximately the same Average Well Colour Development (AWCD) (mean of the absorbance values for an entire plate), for example 0.25, 0.5, 0.75, and 1.00, regardless of the actual incubation time that is required to reach this AWCD for specific plates to be determined. Due to the fact that at this value, the response of a microbial community can be observed in the majority of wells, and the wells that contain the most active microbial communities reach the asymptote of colour development, it is considered to be the ideal solution to choose measurements of AWCD equal to 0.75 (Heuer and Small, 1997; Smalla et al., 1998; Garland and Mills, 1991; Fang et al., 2005; Garland, 1996; Grządziel et al., 2019).

Additionally, kinetic examination of the evolution of colour on plates has been utilized often. The quantity of information that may be gained from a physiological profile is significantly increased by this type of analysis which is being performed. When it comes to the majority of situations, a sigmoidal curve is able to depict the rate of substrate usage (colour development). It is possible to generate a curve that is characterized with three parameters of the reaction kinetics by performing a lot of measurements in a very sequential order (Stefanowicz, 2006; Garland, 1997; Mondini and Insam, 2003; Lindstrom et al., 1998; Nemergut et al., 2013; Verschuere et al., 1997). The lag time, denoted by the symbol λ, is the amount of time that passes between the inoculation of a soil solution into wells and the onset of colour development. The maximum rate of colour development, denoted as μm or r, is a measure that is reflected by the slope of a mathematical curve. The maximum absorbance in a well is denoted by the letters A or K, which stand for threshold or asymptote. Colour development can be depicted in individual wells (separately for various substrates) or in the mean colour development for an entire plate (all substrates) using graphs of this sort (absorbance against time), as shown in Figure 1.

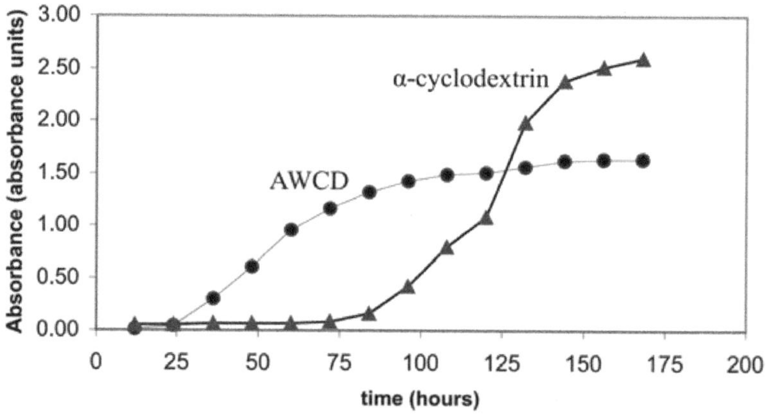

Figure 1. An example of a sigmoidal curve depicting the progression of colour development over 168 hours of incubation for the substrate α-cyclodextrine and AWCD (the average for all substrates on a GN2 plate) (Source: Stefanowicz, 2006).

Through the process of fitting the Gompertz equation to the data, the parameters can be obtained.

$$y = A \exp\left\{ -\exp\left[\frac{\mu_m e}{A}(\lambda - t) + 1 \right] \right\}$$

where y is the absorbance in time t.

Given that the parameters of the Gompertz model have only a weak correlation with one another, it may be concluded that each of these parameters offers a distinct kind of information regarding microbial communities. The utilization of multidimensional statistical approaches is typically required, regardless of the type of absorbance measurement that is selected, which may include the net Optical Density (OD) of individual substrates, the absorption wall thickness (AWCD), or parameters obtained from the kinetic analysis, such as λ, μm, or A. The Principal Component Analysis (PCA) test is the one that is most widely used. As the absorbance of each substrate on a plate (95 substrates in the case of GN and GP plates) or each parameter of each kinetic curve is a variable, the number of variables can be greater than the number of samples. Additionally, the number of samples can be greater than the number of variables. Using Principal Component Analysis (PCA), the number of variables is typically reduced to a few or slightly more than a dozen important Principal Components (PCs). In addition to explaining a piece of the total variability, each principal component (PC) represents a unique mix of starting variables. Then, using correlation analysis of starting variables with PCs, one is able to determine which original variables were responsible for the construction of particular PCs. The data from the Biolog plates are compared using a variety of methodologies, including principal component analysis (PCA), cluster analysis, and Canonical Variate Analysis (CVA), which are used rather frequently (Preston-Mafham et al., 2002; Garland et al., 2001; Stefanowicz, 2006).

BIOLOG plates

Microbial identification using microplates

With microplates assistance, it is possible to identify about 3,000 different species of aerobic and anaerobic bacteria, yeast, and fungi in a quick and accurate manner. Furthermore, the cutting-edge phenotypic technology that Biolog possesses not only makes it possible to identify strains at the species level, but it also provides valuable insights into the functional properties of strains. Through the utilization of discrete test reactions that were carried out in a microplate with 96 wells (Figure 2), Biolog's carbon source utilization method creates a distinctive pattern, or "metabolic fingerprint," that is used to identify pathogenic and environmental bacteria. After testing culture suspensions using a panel of pre-selected assays, they are incubated, read, and cross-referenced with large database (https://www.biolog.com/products/microbial-identification-microplates/).

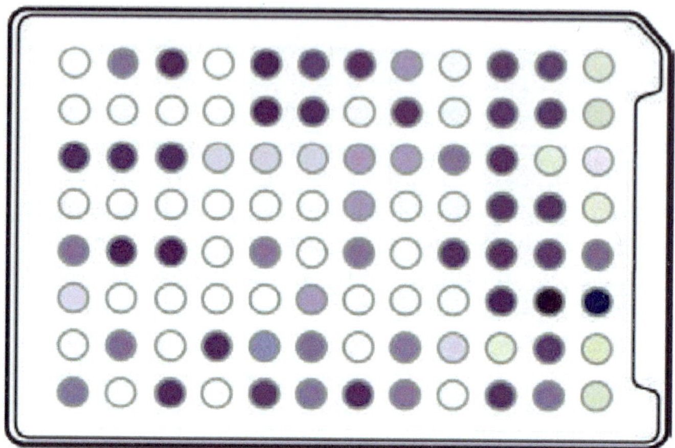

Figure 2. A typical BIOLOG plate. Increasing purple colour intensity represent high activity (Adopted from BIOLOG, 2024 website).

GEN III Test Panel

A wide variety of gram-positive and gram-negative bacteria can be profiled and identified using 94 distinct biochemical assays by the Microbial Identification Test Panel (GEN III Microplate).

Features of GEN III Test Panel

1. Examine a mystery microbe using 94 concurrent phenotypic assays. The microorganism's "Phenotypic Fingerprint," which is used to identify it at the species or strain level, is provided by the test panel.

2. Prefilled and dried 96-well microplates contain all required nutrients and biochemicals. A sensitive test can screen for the use of carbon sources or resistance to inhibitory substances using redox dyes with colourimetric markers.

3. After cultivating your isolate on agar, suspend it in our inoculation solution. After transferring it, incubate it in the GEN III Microplate. The dye will turn purple if the cells are able to use the special substrate or if they are resistant to the inhibitor chemicals in each well. Purple wells' phenotypic fingerprint is compared to Biolog's large database throughout incubation. When a match is established, the isolate is identified at the species or strain level.

Anaerobic Bacteria Test Panel

A quick and comprehensive identification and characterization tool for anaerobic bacteria is the Anaerobe Identification Test Panel (AN Microplate).

1. Evaluate a microorganism that is unknown by doing concurrent phenotypic tests. The microbe can be identified at the species level using its "Phenotypic Fingerprint," which is provided by the test panel.

2. Prefilled and dried 96-well microplates contain all required nutrients and biochemicals. A sensitive test for carbon source consumption is made possible using redox dyes with colourimetric markers.

3. After cultivating the isolate on agar, suspend it in our inoculation solution. Place it in an environment devoid of hydrogen (AN Microplate) and allow it to incubate. The dye will change colour if the cells are able to employ the special substrate in each well by respiring. The phenotypic fingerprint of each well is compared to the vast database of Biolog throughout the incubation period. When a match is discovered, the isolate is identified at the species level.

The Yeast Identification Test Panel

The Yeast Identification Test Panel (YT Microplate) employs 94 biochemical assays to identify and describe a wide variety of yeast species.

1. Use 94 concurrent phenotypic assays to examine an unidentified microbe. To identify the yeast at the species level, the test panel offers a "Phenotypic Fingerprint" of the organism.

2. The 96-well microplate has been prefilled with the required nutrients and biochemicals and dried. The use of redox dyes with colourimetric markers allows for a sensitive analysis for the usage of carbon sources.

3. After yeast has grown, suspend it in our inoculation fluid. Place it in the YT Microplate and let it sit there. The dye will turn purple as a result of respiration by the cells if they are able to employ the special substrate in each well. Purple wells' phenotypic fingerprint is compared to a vast database maintained by Biolog during the incubation period. Finding a match results in an identification of the isolate at the species level.

Filamentous Fungi Test Panel

A quick, all-purpose tool for filamentous fungi and yeast identification and characterization is the Fungal Identification Test Panel (FF Microplate).

1. Using 95 different phenotypic assays at the same time, investigate an unidentified microbe. "Phenotypic Fingerprint" of the microorganism is provided by the test panel, which is utilized for the purpose of identifying the microorganism at the species level.

2. The microplate with 96 wells has been prefilled with all of the essential nutrients and biochemicals, and then it has been dried. The utilization of carbon sources or resistance to inhibitory compounds can be determined with a high degree of precision using redox dyes that are equipped with colourimetric indicators.

3. The isolate should be grown on agar medium, and it should be suspended in our inoculating fluid thereafter. Proceed to incubate it after transferring it on the FF Microplate. The cells will undergo respiration and cause the dye to change colour to a purple hue if they are able to make use of the specific substrate that is present in each well. A comparison is made between the phenotypic fingerprint of purple wells and the vast database that Biolog maintains while the wells are incubating. The identification of the isolate at the species level is carried out once a match has been discovered (BIOLOG, 2024).

Microbial community analysis with EcoPlates

Numerous microorganisms can be found in practically every environment, and they are often the first species to respond to changes in both the chemical and physical atmosphere. Alterations in the populations of microorganisms are frequently a forerunner to alterations in the overall health and viability of the environment. Community-Level Physiological Profiling, also known as CLPP, has been shown to be an effective method for discriminating between changes in microbial communities during different time periods and locations. In the field of applied ecological research, EcoPlatesTM are utilized not only as a method for determining the stability of a normal microbial population, but also for identifying and evaluating changes that occur after the introduction of an environmental variable. The population changes in the soil should be analyzed, water and wastewater should be tested, the effectiveness of bioremediation should be determined, and activated sludge, compost, and industrial waste should be tested (BIOLOG, 2024).

Acknowledgement

We duly acknowledge contents adopted as per CC BY 4.0 from BIOLOG website (https://www.biolog.com/).

References

Arrigo, N. C. and Lipkin, W. I. (2012). Microbe hunting and pathogen discovery. *In*: Institute of Medicine (US). Improving Food Safety Through a One Health Approach: Workshop Summary. Washington (DC): National Academies Press (US); 2012. A10. Available from: https://www.ncbi.nlm.nih.gov/books/NBK114503/.

BIOLOG. Biolog website. https://www.biolog.com/ Assessed on 20-Jan-2024.

Checcucci, A., Luise, D., Modesto, M., Correa, F., Bosi, P., Mattarelli, P. et al. (2021). Assessment of Biolog EcoplateTM method for functional metabolic diversity of aerotolerant pig fecal microbiota. Applied Microbiology and Biotechnology 105(14-15): 6033–6045. https://doi.org/10.1007/s00253-021-11449-x.

Fang, M., Kremer, R. J., Motavalli, P. P. and Davis, G. (2005). Bacterial diversity in rhizospheres of nontransgenic and transgenic corn. Applied and Environmental Microbiology 71(7): 4132–4136. https://doi.org/10.1128/AEM.71.7.4132-4136.2005.

Ferrer, J., Prats, C. and López, D. (2008). Individual-based modelling: an essential tool for microbiology. Journal of Biological Physics 34(1-2): 19–37. https://doi.org/10.1007/s10867-008-9082-3.

Garland, J. L. (1966). Analytical approaches to the characterization of samples of microbial communities using patterns of potential c utilization. Soil Biol. Biochem. 28: 213–221. http://dx.doi.org/10.1016/0038-0717(95)00112-3.

Garland, J. L. (1997). Analysis and interpretation of community level physiological profiles in microbial ecology. FEMS Microbiol. Ecol. 24: 289.

Garland, J. L. and Mills, A. L. (1991). Classification and characterization of heterotrophic microbial communities on the basis of patterns of community-level sole-carbon-source utilization. Applied and Environmental Microbiology 57(8): 2351–2359. https://doi.org/10.1128/aem.57.8.2351-2359.1991.

Garland, J. L., Mills, A. L. and Young, J. S. (2001). Relative effectiveness of kinetic analysis vs single point readings for classifying environmental samples based on community-level physiological profiles (CLPP). Soil Biol. Biochem. 33: 1059.

Gibbons, S. M. and Gilbert, J. A. (2015). Microbial diversity—exploration of natural ecosystems and microbiomes. Current Opinion in Genetics & Development 35: 66–72. https://doi.org/10.1016/j.gde.2015.10.003.

Gryta, A., Frąc, M. and Oszust, K. (2014). The application of the Biolog EcoPlate approach in ecotoxicological evaluation of dairy sewage sludge. Applied Biochemistry and Biotechnology 174(4): 1434–1443. https://doi.org/10.1007/s12010-014-1131-8.

Grządziel, J., Furtak, K. and Gałązka, A. (2019). Community-level physiological profiles of microorganisms from different types of soil that are characteristic to Poland—a long-term microplot experiment. Sustainability 11: 56. https://doi.org/10.3390/su11010056.

Heuer, H. and Small, K. (1997). Evaluation of community-level catabolic profiling using BIOLOG GN microplates to study microbial community changes in potato phyllosphere. J. Microbiol. Methods 30(1): 49–61. https://doi.org/10.1016/S0167-7012(97)00044-4.

Hitzl, W., Rangger, A., Sharma, S. and Insam, H. (1997). Separation power of the 95 substrates of the BIOLOG system determined in various soils. FEMS Microbiology Ecology 22(3): 167–174. https://doi.org/10.1111/j.1574-6941.1997.tb00368.x.

Huang, Z. (2023). The hidden world of soil microbes: Guardians of ecosystem health. Ukrainian Journal of Ecology 13(7): 28–30. doi: 10.15421/2023_477.

Jeswani, H. K., Chilvers, A. and Azapagic, A. (2020). Environmental sustainability of biofuels: a review. Proceedings. Mathematical, Physical, and Engineering Sciences 476(2243): 20200351. https://doi.org/10.1098/rspa.2020.0351.

Kapinusova, G., Lopez Marin, M. A. and Uhlik, O. (2023). Reaching unreachables: Obstacles and successes of microbial cultivation and their reasons. Frontiers in Microbiology 14: 1089630. https://doi.org/10.3389/fmicb.2023.1089630.

Kelly, J. J. and Tate, R. L. (1998). Use of BIOLOG for the analysis of microbial communities from zinc-contaminated soils. J. Environ. Quality 27(3): 600. https://doi.org/10.2134/jeq1998.00472425002700030018x.

Lane, N. (2015). The unseen world: reflections on Leeuwenhoek (1677) Concerning little animals. Philosophical Transactions of the Royal Society of London. Series B, Biological Sciences 370(1666): 20140344. https://doi.org/10.1098/rstb.2014.0344.

Lim, J., Wehmeyer, H., Heffner, T., Aeppli, M., Gu, W., Kim, P. J. et al. (2024). Resilience of aerobic methanotrophs in soils; spotlight on the methane sink under agriculture. FEMS Microbiology Ecology 100(3): fiae008. https://doi.org/10.1093/femsec/fiae008.

Lindstrom, J. E., Barry, R. P. and Braddock, J. F. (1998). Microbial community analysis: a kinetic approach to constructing potential C source utilization patterns. Soil Biol. Biochem. 30(2): 231–239. https://doi.org/10.1016/S0038-0717(97)00113-2.

Lipkin, W. I. (2010). Microbe hunting. Microbiology and Molecular Biology Reviews : MMBR 74(3): 363–377. https://doi.org/10.1128/MMBR.00007-10.

Mondini, C. and Insam, H. (2003). Community level physiological profiling as a tool to evaluate compost maturity: a kinetic approach. Eur. J. Soil Biol. 39: 141.

Németh, I., Molnár, S., Vaszita, E. and Molnár, M. (2021). The Biolog EcoPlate™ technique for assessing the effect of metal oxide nanoparticles on freshwater microbial communities. Nanomaterials (Basel, Switzerland) 11(7): 1777. https://doi.org/10.3390/nano11071777.

Nemergut, D. R., Schmidt, S. K., Fukami, T., O'Neill, S. P., Bilinski, T. M., Stanish, L. F. et al. (2013). Patterns and processes of microbial community assembly. Microbiology and Molecular Biology Reviews: MMBR 77(3): 342–356. https://doi.org/10.1128/MMBR.00051-12.

Pierce, M. L., Ward, J. E. and Dobbs, F. C. (2014). False positives in Biolog EcoPlates™ and MT2 MicroPlates™ caused by calcium. Journal of Microbiological Methods 97: 20–24. https://doi.org/10.1016/j.mimet.2013.12.002.

Preston-Mafham, J., Boddy L. and Randerson P. F. (2002). Analysis of microbial community functional diversity using sole-carbon-source utilisation profiles—a critique. FEMS Microbiol. Ecol. 42: 1.

Smalla, K., Wachtendorf, U., Heuer, H., Liu, W. T. and Forney, L. (1998). Analysis of BIOLOG GN substrate utilization patterns by microbial communities. Applied and Environmental Microbiology 64(4): 1220–1225. https://doi.org/10.1128/AEM.64.4.1220-1225.1998.

Smith, H. (1984). The biochemical challenge of microbial pathogenicity. The Journal of Applied Bacteriology 57(3): 395–404. https://doi.org/10.1111/j.1365-2672.1984.tb01405.x.

Sohn, S. I., Ahn, J. H., Pandian, S., Oh, Y. J., Shin, E. K., Kang, H. J. et al. (2021). Dynamics of bacterial community structure in the rhizosphere and root nodule of soybean: impacts of growth stages and varieties. International Journal of Molecular Sciences 22(11): 5577. https://doi.org/10.3390/ijms22115577.

Stefanowicz, A. (2006). The Biolog plates technique as a tool in ecological studies of microbial communities. Polish J. of Environ. Stud. 15(5): 669–676.

Tao, X., Yang, Z., Feng, J., Jian, S., Yang, Y., Bates, C. T. et al. (2024). Experimental warming accelerates positive soil priming in a temperate grassland ecosystem. Nature Communications 15(1): 1178. https://doi.org/10.1038/s41467-024-45277-0.

The Microbial World: Foundation of the Biosphere: This report is based on an American Academy of Microbiology colloquium held January 19–21, 1996, in Palm Coast, Florida. The colloquium was supported by the National Science Foundation, the National Oceanic and Atmospheric Administration of the U.S. Department of Commerce, the U.S. Department of Energy, and the American Society for Microbiology. Washington (DC): American Society for Microbiology; 1997. Available from: https://www.ncbi.nlm.nih.gov/books/NBK562919/ doi: 10.1128/AAMCol.19Jan.1996.

van der Heijden, M. G., Bardgett, R. D. and van Straalen, N. M. (2008). The unseen majority: soil microbes as drivers of plant diversity and productivity in terrestrial ecosystems. Ecology Letters 11(3): 296–310. https://doi.org/10.1111/j.1461-0248.2007.01139.x.

Verschuere, L., Fever, V., van Vooren, L. and Vertstraete, W. 1997. The contribution of individual population to the Biolog pattern of model microbial communities. FEMS Microbiology Ecol. 24: 353–362.

Weber, K. P., Grove, J. A., Gehder, M., Anderson, W. A. and Legge, R. L. (2007). Data transformations in the analysis of community-level substrate utilization data from microplates. Journal of Microbiological Methods 69(3): 461–469. https://doi.org/10.1016/j.mimet.2007.02.013.

Zhang, W., Han, J., Wu, H., Zhong, Q., Liu, W., He, S. et al. (2021). Diversity patterns and drivers of soil microbial communities in urban and suburban park soils of Shanghai, China. PeerJ 9: e11231. https://doi.org/10.7717/peerj.11231.

Chapter 4
Overview of DNA Library Preparation Techniques for Next-Generation Sequencing

Introduction

Next-Generation Sequencing (NGS) is a technology that has changed our ability to study and comprehend a wide variety of biological concerns. For this reason, DNA library preparation is an essential step in the process. In order for a DNA sample to be deciphered by a next-generation sequencing (NGS) machine, it must first be subjected to rigorous processing in order to produce a DNA library (Radford et al., 2012; Behjati and Tarpey, 2013; Bahassi and Stambrook, 2014; Qin, 2019; Gupta and Verma, 2019; Cheng et al., 2023; Akintunde et al., 2023; Satam et al., 2023). Let us break down what this process entails and why it's so crucial. Imagine that DNA is a massive encyclopedia that is packed with information about genetic details. In order to prepare the DNA library, the first step is DNA fragmentation. In DNA fragmentation, the DNA strand is broken into substantial strands and more manageable pieces. As next-generation sequencing systems are designed to read small snippets rather than complete chromosomes, this is an extremely important consideration. Enzymatically, which involves the utilization of particular enzymes, or mechanically, which involves the utilization of techniques such as sonication, which makes use of sound waves, are both viable options for fragmentation. The next procedure is analogous to fixing any edges of our encyclopedia pages that have become frayed or damaged which could have occurred. Ensuring that the ends of the DNA fragments are smooth and uniform is the purpose of this procedure, which is known as end repair. Additionally, this is essential for the subsequent phase, which involves the attachment of specialized adapters. Short segments of DNA known to have specific sequences are called adapters, and they act as "handles" for the NGS apparatus. Additionally, they carry one-of-a-kind barcodes, which enable us to keep track of distinct samples even while we are analyzing several samples inside the same sequencing run. Primers are used for sequencing reactions. Lignation is the term used to describe the process of binding these adapters to the DNA fragments. It is possible that the quantity of DNA that has been produced is not always sufficient to produce reliable sequencing findings. PCR amplification comes into play in this particular scenario. Your DNA fragments are effectively copied several times during the Polymerase Chain Reaction (PCR), which results in a larger DNA library that can be read by the sequencer. The last step is a cleanup process, which entails eliminating any residual pieces that might interfere with the sequencing reactions at some point. Now, following the methods given above, the DNA library is prepared to be used by the NGS equipment (Head et al., 2014; Gupta and Verma, 2019; Gansauge and Meyer, 2019; Dolgova et al., 2020; van Dijk et al., 2014; Loudig et al., 2018; Bronner and Quail, 2019).

Why is DNA library preparation so important and necessary?

DNA library preparation is needed for NGS because it changes the form of raw DNA data so that they can work with the chosen NGS platform. To do this, the DNA has to be broken up into pieces that can be handled, adapters have to be added so that the pieces can stick to the sequencing flow cell and be amplified, and barcode sequences might need to be added so that different samples can be read at the same time. Without proper library preparation, the NGS sequencer would be unable to accurately read and interpret the genetic information contained within the sample, rendering the sequencing process ineffective (Head et al., 2014; Haendiges et al., 2021).

Types of library preparation technique

There are two main types of library preparation techniques used for next-generation sequencing (NGS) such as (a) Ligation-based library preparation and (b) Tagmentation-based library preparation.

a) *Ligation-based library preparation* (Kapp et al., 2021; Cheng et al., 2022; Miura et al., 2023)

This is the traditional and more established method. Here is a breakdown of the steps involved:

- *Fragmentation*: Similar to both techniques, DNA is first fragmented into smaller pieces using enzymes or mechanical shearing. The chosen enzyme or method depends on the desired fragment size and the specific application.

- *End repair*: The fragmented DNA ends are blunt-ended using enzymes to ensure efficient ligation with adapters in the next step.

- *Adapter ligation*: Short adapter sequences are ligated (attached) to the blunt ends of the DNA fragments. These adapters serve two key purposes: They contain sequences that are complementary to the flow cell surface in the NGS instrument, allowing the fragments to bind for sequencing. They may contain unique barcode sequences, allowing researchers to multiplex (sequence multiple samples) in a single sequencing run. These barcodes are then used to differentiate reads originating from different samples during data analysis.

- *Optional PCR amplification*: If the starting amount of DNA is low, PCR amplification can be used to increase the number of DNA fragments in the library. This ensures sufficient material for sequencing and improves the accuracy of the results.

- *Library purification*: The final step involves purifying the library to remove any leftover contaminants or unused adapters that could interfere with the sequencing process.

b) *Tagmentation-based library preparation* (Adey, 2021; Weichenhan et al., 2018; Ribarska et al., 2022; Gustafsson et al., 2023)

This is a faster and more streamlined method compared to ligation-based techniques. Here is what makes it different:

- *Fragmentation and adapter attachment (combined)*: This is the key step that differentiates tagmentation from ligation-based methods. Here, a transposase enzyme is used. This enzyme performs two critical functions simultaneously:
  - *DNA fragmentation*: The transposase (Hickman and Dyda, 2015; Kia et al., 2017; Tanaka et al., 2020) cuts the DNA strands at specific recognition sequences, effectively fragmenting the DNA into smaller pieces.

    ○ *Adapter attachment*: While fragmenting the DNA, the transposase also attaches adapter sequences to the ends of the fragments. These adapter sequences serve similar purposes as in ligation-based methods: They contain sequences complementary to the flow cell surface in the NGS instrument, allowing the fragments to bind for sequencing. They may contain unique barcode sequences, allowing researchers to multiplex (sequence multiple samples) in a single sequencing run and differentiate them later during data analysis.

- *Optional PCR amplification*: Similar to ligation-based methods, PCR amplification may be used if the starting amount of DNA is low. This step amplifies the DNA fragments, increasing the number of copies in the library and ensuring sufficient material for accurate sequencing.

- *Library purification*: The final step involves purifying the library to remove any leftover contaminants or unused adapters that could interfere with the subsequent sequencing process.

Choosing the Right Technique:

The choice between ligation-based and tagmentation-based library preparation depends on several factors:

Sample type and quality: Ligation-based methods offer more flexibility for handling damaged DNA samples.

Desired fragment size: Some experiments require very specific fragment sizes, which may be easier to achieve with enzymatic fragmentation in the ligation-based method.

Throughput: Tagmentation-based methods are generally faster, making them ideal for high-throughput NGS projects.

Cost: Ligation-based methods typically require more enzymatic steps and may be slightly more expensive.

Both library preparation techniques have their advantages and disadvantages. Understanding these nuances allows researchers to select the most appropriate method for their specific NGS experiment (Head et al., 2014; Zhao et al., 2020; Gaulke et al., 2021).

Steps for library preparation

Kits for preparing libraries use a variety of approaches. The next-generation sequencing (NGS) library prep kits optimized for Illumina sequencers allow you to ask nearly any question relating to the genome, transcriptome, or epigenome of any organism. These kits come in a variety of formats. Innovative, user-friendly, and rapid methods for the production of DNA and RNA libraries. Kits that have been developed specifically for our instruments and secondary analysis platform will provide you access to improved support throughout the entire work flow. These are listed as follows:

1. *AmpliSeq for Illumina BRCA Panel*: Targeted research panel examining somatic and germline variants in BRCA1 and BRCA2.

2. *AmpliSeq for Illumina Cancer Hotspot Panel v2*: Targeted research panel examining hotspot regions of 50 genes with known associations to cancer.

3. *AmpliSeq for Illumina Childhood Cancer Panel*: Targeted panel for investigating 203 genes associated with cancer in children and young people.

4. *AmpliSeq for Illumina Comprehensive Panel v3*: Targeted DNA and RNA research panel investigating variants across 161 genes related with a range of cancer types.

5. *AmpliSeq for Illumina Custom DNA Panel*: Targeted custom study panels optimized for sequencing specific targets or genomic content of interest.

6. *AmpliSeq for Illumina Custom RNA Fusion Panel*: A customizable targeted panel for finding fusion genes and measuring gene expression that can include up to 1200 targets of interest.

7. *AmpliSeq for Illumina Custom RNA Panel*: Targeted, custom RNA research panels designed for sequencing up to 1200 genomic targets of interest.

8. *AmpliSeq for Illumina Focus Panel*: Targeted DNA and RNA research panel investigating 52 genes with known relation to solid tumours.

9. *AmpliSeq for Illumina Immune Repertoire Plus, TCR beta Panel*: Targeted RNA research panel to explore T cell diversity and clonal expansion by sequencing T cell receptor beta chain rearrangements.

10. *AmpliSeq for Illumina Immune Response Panel*: Targeted RNA expression panel studying 395 genes involved in tumour-immune system interactions.

11. *AmpliSeq for Illumina Library Prep, Indexes, and Accessories*: Rapidly prepare amplicon libraries for Illumina sequencers using AmpliSeq for Illumina library preparation and index adapters. AmpliSeq for Illumina Myeloid Panel: Targeted panel to study 40 DNA genes, 29 RNA fusion driver genes, and 5 gene expression levels associated with myeloid cancers.

12. *AmpliSeq for Illumina On-Demand*: Tailor panel designs for human disease study efficiently and conveniently by selecting from a catalogue of over 5,000 pretested genes.

13. *AmpliSeq for Illumina TCR beta-SR Panel*: FFPE-compatible panel for measuring T cell diversity and clonal growth in tumour samples by sequencing T cell receptor beta chain rearrangements.

14. *AmpliSeq for Illumina Transcriptome Human Gene Expression Panel*: Targeted panel that measures expression levels of > 20,000 human RefSeq genes.

15. *Assay for COVIDSeq (96 samples)*: This low- to medium-throughput NGS assay lets laboratories find new types of SARS-CoV-2 and keep track of how common they are.

16. The COVIDSeq Test (RUO Version) is a high-throughput next-generation sequencing (NGS) test that finds changes in the SARS-CoV-2 virus.

17. Illumina Complete Long Read Prep, Human is a full workflow option for accurate and fast human WGS using long-read data from NovaSeq platforms.

18. Illumina DNA PCR-Free Prep is a fast, high-performance process that can be used for sensitive tasks like human whole-genome sequencing.

19. There is a quick, unified process called Illumina DNA Prep that can be used for many things, from reading the whole human genome to working with amplicons, plasmids, and microbes.

20. *DNA Prep with Enrichment from Illumina*: A quick, unified process that can be used for many different tasks and offers tailored resequencing for fixed panels, custom panels, and whole-exome enrichment.

21. Illumina DNA Prep with Exome 2.5 Enrichment is a complete, high-performance, and quick whole-exome sequencing kit that comes with indexes, clean up/size selection beads, a full exome probe panel, and library prep and hybridization chemicals.

22. The Illumina Microbial Amplicon Library Prep is a flexible and streamlined NGS library prep kit that can be used for a number of public health surveillance and microbial study tasks.

23. The Illumina Microbial Amplicon Prep—Influenza A/B kit is an easy-to-use whole-genome NGS library prep kit for useful study and public health monitoring of influenza A and B viruses.

24. With Illumina RNA Prep with Enrichment, you can quickly and accurately test a large group of target genes. This is made possible by its high capture efficiency and coverage regularity.

25. It is very good at finding and identifying common respiratory viruses, like SARS-CoV-2 types, and the Illumina Respiratory Virus Enrichment Kit does this quickly and completely.

26. Rapid library preparation from a wide range of sample types for studying the coding and non-coding transcriptome with a lot of study freedom. This is possible with Illumina Stranded Total RNA Prep with Ribo-Zero Plus or Ribo-Zero Plus Microbiome.

27. Illumina Stranded mRNA Prep is an easy, scalable, cost-effective, and fast way to analyze the coding transcriptome in just one day, using as little as 25 ng of normal (non-degraded) RNA as input.

28. Multiplex panel, library prep, indexes, and bioinformatic analysis for NGS sequencing of mycobacterium, M. tuberculosis complex (MTBC) strains, and antibiotic resistance markers are all part of the Illumina and GenoScreen Deeplex® Myc-TB Combo Kit.

29. With the Nextera XT DNA Library Preparation Kit, you can make sequencing libraries for small genomes, PCR amplicons, plasmids, or cDNA in just 90 minutes, and you only need a small amount of DNA.

30. *Pan-Coronavirus Panel*: This panel lets you describe at least 200 known coronavirus families so that you can keep an eye on animal reservoirs and sequence known coronaviruses.

31. *Pillar oncoReveal Multi-Cancer CNV + RNA Fusion Panel*: This is a 78-gene pan-cancer panel that has both DNA and RNA content. It uses the single-tube SLIMampTM process to make libraries quickly and get them ready for use.

32. The Pillar® oncoRevealTM BRCA1 & BRCA2 + CNV Panel is a targeted BRCA1/2 gene panel that uses the single-tube SLIMampTM process to make libraries quickly and get them ready for use.

33. Assay that targets key mutations in the MPL, JAK2, and CALR genes using a single-tube SLIMamp workflow for fast library preparation and response time. This is the Pillar® oncoRevealTM Essential MPN Panel | Oncology targeted NGS panel.

34. The Pillar® oncoRevealTM Myeloid Panel is an oncology-focused NGS panel that includes 78 genes related to pan-cancer. It has both DNA and RNA content and uses the single-tube SLIMampTM process for quick library preparation and turnaround time.

35. AMR Enrichment Panel Kit for Respiratory Pathogens.

36. This NGS panel looks for more than 280 respiratory pathogens, such as SARS-CoV-2, and more than 2000 antimicrobial resistance genes. Explify RPIP is used to analyze the data.

37. This is the Ribo-Zero Plus Microbiome rRNA Depletion Kit. It gets rid of unwanted host and pan-bacterial rRNA quickly and effectively from complex microbial samples (like stool) for metatranscriptomics study.

38. The TruSeq ChIP Library Preparation Kit makes it easy and inexpensive to prepare a chromatin immunoprecipitation sequencing (ChIP-Seq) DNA library. It comes with master mixes and strong multiplexing capabilities.

39. The TruSeq DNA Exome kit is a cheap way to sequence the entire genome and offers excellent target coverage across a wide range of read depths.

40. TruSeq DNA Nano lets you make whole-genome sequencing libraries and quickly look into samples that don't have a lot of DNA.

41. *Without PCR, TruSeq DNA*: A simple, all-inclusive method for making a library for Whole-Genome Sequencing (WGS) that covers complex genomes completely and accurately.

42. *TruSeq RNA Exome*: It is a cheap and reliable way to sequence RNA from FFPE tissues and other low-quality materials. It works with as little as 10 ng of total RNA.

43. *TruSeq RNA Library Prep Kit v2*: These kits make it easy and cheap to analyze the coding transcriptome with little time spent doing the work.

44. *How to Make a TruSeq Small RNA Library*: Making miRNA and small RNA sequencing libraries straight from total RNA for any species is easy and does not cost much with these kits.

45. TruSeq Stranded Total RNA is a strong and scalable whole-transcriptome analysis (RNA-Seq) system that works with a wide range of species and sample types, such as human, mouse, and Formalin-Fixed, Paraffin-Embedded (FFPE) tissue.

46. With TruSeq Stranded Total RNA with Ribo-Zero Globin, you can make whole-transcriptome sequencing libraries from RNA taken from blood in a quick and easy way that gets rid of ribosomal RNA and globin mRNA at the same time.

47. *TruSeq Stranded Total RNA with Ribo-Zero Plant*: This RNA-Seq library prep process lets you get a full picture of the plant transcriptome. It gets rid of the rRNA in the plant's leaves, seeds, and roots.

48. Prepare sequencing libraries from mRNA with TruSeq Stranded mRNA to get a clear picture of the coding transcriptome and get information that is specific to each strand.

49. The TruSight Hereditary Cancer Panel includes material that was chosen by experts to focus on 113 genes that are linked to a higher risk of getting cancer.

50. The TruSight Oncology 500 test looks for a number of different types of variants, such as microsatellite instability (MSI) and Tumour Mutational Burden (TMB).

51. The TruSight Oncology 500 High-Throughput makes it possible to do full genome profiling on FFPE samples and lets you batch up to 192 samples per flow cell.

52. The TruSight Oncology 500 ctDNA test looks for a number of different types of somatic variants in plasma, such as microsatellite instability (MSI) and tumour mutational burden (TMB).

53. The TruSight RNA Fusion Panel can find all kinds of gene fusions in formalin-fixed, paraffin-embedded (FFPE) and other types of cancer study samples.

54. Using TruSight RNA Pan-Cancer to look for 1385 cancer genes for gene expression, variation, and fusion studies in a variety of

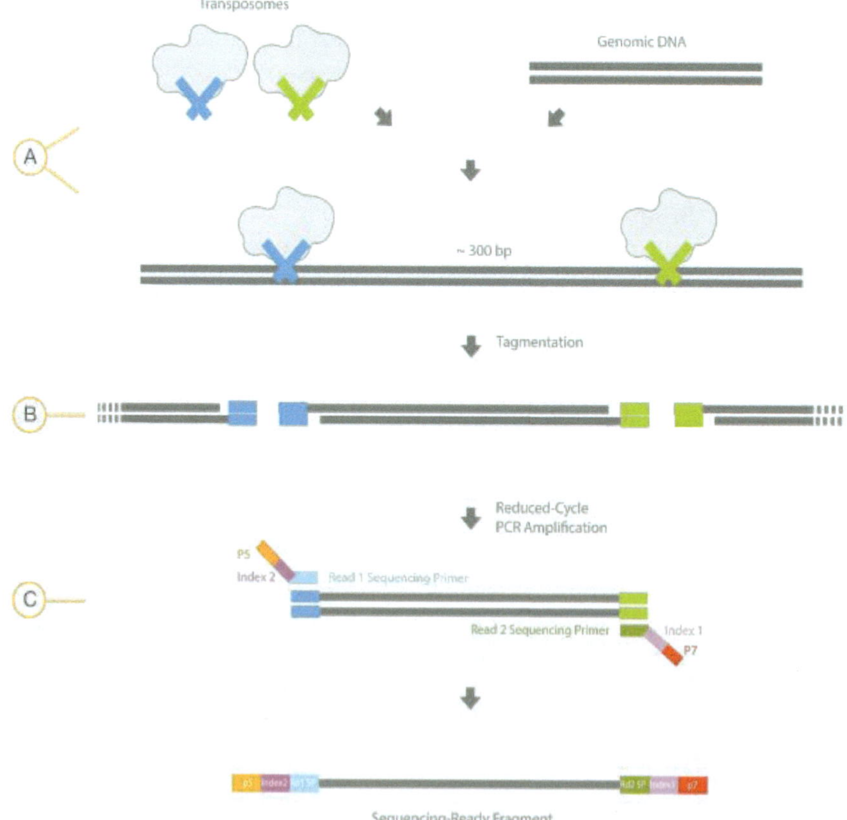

Figure 1. Overview of Nextera XT DNA Library Prep workflow (Source: Illumina, 2023a).

RNA samples, such as blood, bone marrow, and formalin-fixed, paraffin-embedded (FFPE) tissue.

55. TruSight Tumour 15 is a focused sequencing panel that checks 15 genes that are often changed in solid tumours in a single test that can be done quickly and easily.

56. TruSight Tumour 170 is a full next-generation sequencing (NGS) test that looks for DNA and RNA variations in the same FFPE sample.

57. *Urinary Pathogen ID/AMR Enrichment Kit*: The Urinary Pathogen ID/AMR Panel identifies and counts common and less well known uropathogens in a culture-free way that is very precise.

58. *Viral Surveillance Panel*: The Viral Surveillance Panel gives target enrichment and whole-genome sequencing and characterization of 66 viruses, such as SARS-CoV-2, influenza, and polio.

Considering the huge number of kits available and continuously developing new kits and specific panels. It it very difficult to stick to one protocol. Hence, it is necessary to check regular developments and use new and existing protocols for library preparation. Of these, Nextera XT DNA Library Prep is one of kit primarily used for prepare up to 384 dual-indexed paired-end libraries from DNA (Figures 1 and 2) (Illumina,

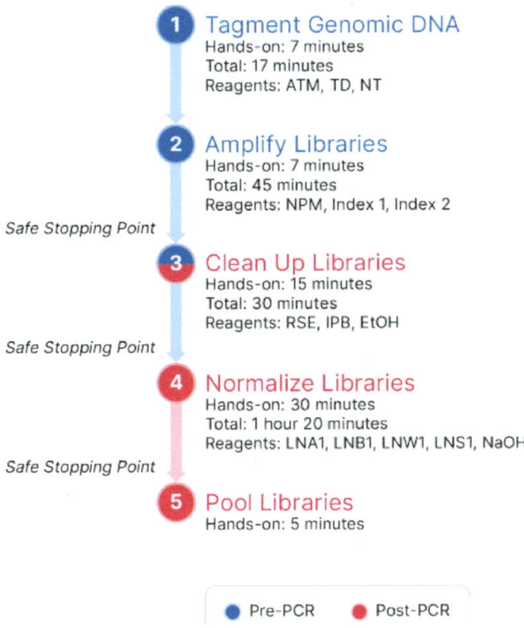

Figure 2. Illustration of the Nextera XT DNA Library Prep workflow (Source: Illumina, 2023b).

2023b). Detailed method for Nextera XT DNA Library Prep can be assessed using link: https://support-docs.illumina.com/LP/NexteraXTRef/Content/LP/Nextera/XT/Overview.htm.

Moreover, refer Illumina DNA Prep Reference Guide for the detailed step-by-step protocol (Illumina, 2023c; Rekadwad, 2023).

Acknowledgement

The author duly acknowledges the website contents adopted as per CC BY 4.0 from Illumina Inc., website (https://www.illumina.com/)

References

Adey, A. C. (2021). Tagmentation-based single-cell genomics. Genome Research 31(10): 1693–1705. https://doi.org/10.1101/gr.275223.121.

Akintunde, O., Tucker, T. and Carabetta, V. J. (2023). The evolution of next-generation sequencing technologies. ArXiv, arXiv:2305.08724v1.

Bahassi, M. and Stambrook, P. J. (2014). Next-generation sequencing technologies: breaking the sound barrier of human genetics. Mutagenesis 29(5): 303–310. https://doi.org/10.1093/mutage/geu031.

Behjati, S. and Tarpey, P. S. (2013). What is next generation sequencing?. Archives of disease in childhood. Education and Practice Edition 98(6): 236–238. https://doi.org/10.1136/archdischild-2013-304340.

Bronner, I. F. and Quail, M. A. (2019). Best practices for Illumina library preparation. Current Protocols in Human Genetics 102(1): e86. https://doi.org/10.1002/cphg.86.

Cheng, C., Fei, Z. and Xiao, P. (2023). Methods to improve the accuracy of next-generation sequencing. Frontiers in Bioengineering and Biotechnology 11: 982111. https://doi.org/10.3389/fbioe.2023.982111.

Cheng, L. Y., Dai, P., Wu, L. R., Patel, A. A. and Zhang, D. Y. (2022). Direct capture and sequencing reveal ultra-short single-stranded DNA in biofluids. iScience 25(10): 105046. https://doi.org/10.1016/j.isci.2022.105046.

Dolgova, A. S., Safonova, M. V. and Dedkov, V. G. (2020). Universal library preparation protocol for efficient high-throughput sequencing of double-stranded RNA viruses. Methods in Molecular Biology (Clifton, N.J.) 2063: 181–188. https://doi.org/10.1007/978-1-0716-0138-9_14.

Gansauge, M. T. and Meyer, M. (2019). A method for single-stranded ancient DNA library preparation. Methods in Molecular Biology (Clifton, N.J.) 1963: 75–83. https://doi.org/10.1007/978-1-4939-9176-1_9.

Gaulke, C. A., Schmeltzer, E. R., Dasenko, M., Tyler, B. M., Vega Thurber, R. and Sharpton, T. J. (2021). Evaluation of the effects of library preparation procedure and sample characteristics on the accuracy of metagenomic profiles. mSystems 6(5): e0044021. https://doi.org/10.1128/mSystems.00440-21.

Gupta, N. and Verma, V. K. (2019). Next-generation sequencing and its application: empowering in public health beyond reality. Microbial Technology for the Welfare of Society 17: 313–341. https://doi.org/10.1007/978-981-13-8844-6_15.

Gustafsson, C., Hauenstein, J., Frengen, N., Krstic, A., Luc, S. and Månsson, R. (2023). T-RHEX-RNAseq—a tagmentation-based, rRNA blocked, random hexamer primed RNAseq method for generating stranded RNAseq libraries directly from very low numbers of lysed cells. BMC Genomics 24(1): 205. https://doi.org/10.1186/s12864-023-09279-4.

Haendiges, J., Jinneman, K. and Gonzalez-Escalona, N. (2021). Choice of library preparation affects sequence quality, genome assembly, and precise *in silico* prediction of virulence genes in shiga toxin-producing *Escherichia coli*. PloS One 16(3): e0242294. https://doi.org/10.1371/journal.pone.0242294.

Head, S. R., Komori, H. K., LaMere, S. A., Whisenant, T., Van Nieuwerburgh, F., Salomon, D. R. et al. (2014). Library construction for next-generation sequencing: overviews and challenges. BioTechniques 56(2): 61–passim. https://doi.org/10.2144/000114133.

Hickman, A. B. and Dyda, F. (2015). Mechanisms of DNA transposition. Microbiology Spectrum 3(2): MDNA3–2014. https://doi.org/10.1128/microbiolspec.MDNA3-0034-2014.

Illumina. (2023a). https://sapac.illumina.com/science/technology/next-generation-sequencing.html (Assessed 05-10-2023).

Illumina. (2023b). Nextera XT DNA Library Prep. https://support-docs.illumina.com/LP/NexteraXTRef/Content/LP/Nextera/XT/Overview.htm (Assessed 05-10-2023).

Illumina. (2023c). Illumina DNA Prep Reference Guide. https://support.illumina.com/content/dam/illumina-support/documents/documentation/chemistry_documentation/illumina_prep/illumina-dna-prep-reference-guide-1000000025416-09.pdf (Assessed 07-10-2023).

Kapp, J. D., Green, R. E. and Shapiro, B. (2021). A fast and efficient single-stranded genomic library preparation method optimized for ancient DNA. Journal of Heredity 112(3): 241–249. https://doi.org/10.1093/jhered/esab012.

Kia, A., Gloeckner, C., Osothprarop, T., Gormley, N., Bomati, E., Stephenson, M. et al. (2017). Improved genome sequencing using an engineered transposase. BMC Biotechnology 17(1): 6. https://doi.org/10.1186/s12896-016-0326-1.

Loudig, O., Liu, C., Rohan, T. and Ben-Dov, I. Z. (2018). Retrospective MicroRNA sequencing: complementary DNA library preparation protocol using formalin-fixed paraffin-embedded RNA specimens. Journal of Visualized Experiments : JoVE (135): 57471. https://doi.org/10.3791/57471.

Miura, F., Kanzawa-Kiriyama, H., Hisano, O., Miura, M., Shibata, Y., Apache, N. et al. (2023). A highly efficient scheme for library preparation from single-stranded DNA. Sci. Rep. 13: 13913. https://doi.org/10.1038/s41598-023-40890-3.

Qin, D. (2019). Next-generation sequencing and its clinical application. Cancer Biology & Medicine 16(1): 4–10. https://doi.org/10.20892/j.issn.2095-3941.2018.0055.

Radford, A. D., Chapman, D., Dixon, L., Chantrey, J., Darby, A. C. and Hall, N. (2012). Application of next-generation sequencing technologies in virology. The Journal of General Virology 93(Pt 9): 1853–1868. https://doi.org/10.1099/vir.0.043182-0.

Rekadwad, B. N. (2023). A reference guide for Illumina Library Preparation. Zenodo. https://doi.org/10.5281/zenodo.8416480.

Ribarska, T., Bjørnstad, P. M., Sundaram, A. Y. M. and Gilfillan, G. D. (2022). Optimization of enzymatic fragmentation is crucial to maximize genome coverage: a comparison of library preparation methods for Illumina sequencing. BMC Genomics 23(1): 92. https://doi.org/10.1186/s12864-022-08316-y.

Satam, H., Joshi, K., Mangrolia, U., Waghoo, S., Zaidi, G., Rawool, S. et al. (2023). Next-generation sequencing technology: current trends and advancements. Biology 12(7): 997. https://doi.org/10.3390/biology12070997.

Tanaka, N., Takahara, A., Hagio, T., Nishiko, R., Kanayama, J., Gotoh, O. et al. (2020). Sequencing artifacts derived from a library preparation method using enzymatic fragmentation. PLoS ONE 15(1): e0227427. https://doi.org/10.1371/journal.pone.0227427.

van Dijk, E. L., Jaszczyszyn, Y. and Thermes, C. (2014). Library preparation methods for next-generation sequencing: tone down the bias. Experimental Cell Research 322(1): 12–20. https://doi.org/10.1016/j.yexcr.2014.01.008.

Weichenhan, D., Wang, Q., Adey, A., Wolf, S., Shendure, J., Eils, R. et al. (2018). Tagmentation-based library preparation for low DNA input whole genome bisulfite sequencing. Methods in Molecular Biology (Clifton, N.J.) 1708: 105–122. https://doi.org/10.1007/978-1-4939-7481-8_6.

Zhao, S., Zhang, C., Mu, J., Zhang, H., Yao, W., Ding, X. et al. (2020). All-in-one sequencing: an improved library preparation method for cost-effective and high-throughput next-generation sequencing. Plant Methods 16: 74. https://doi.org/10.1186/s13007-020-00615-3.

Chapter 5

Bioinformatics Tools for the Identification and Phylogenetic Analysis of Microorganisms

Introduction

Prokaryotic genomes are essential for thorough studies on genetic and functional similarity among microorganisms and give novel insights about new characters in a species (Konstantinidis and Tiedje, 2005). Culturing methods have avoided or failed to achieve 100% isolation of microbial taxa in a pure culture form (Connon and Giovannoni, 2002; Martiny, 2019). In contrast, bulk DNA extracted from samples proved that sequencing and computer analysis tools have considerably improved and are able to recover most of the sequence data in an uncultured way. More than > 80% of microbial taxa are represented by uncultured species (Hugenholtz et al., 2016).

The great majority of the unseen microbial world on Earth needs a close tap to reveal their genetic similarity or diversity, metabolomics profiles, and their potential industrious use (Rosselló-Móra and Whitman, 2019; Romalde et al., 2019). Various software solutions, and service websites become invaluable for "Microbial Systematics" and "Microbe hunting" (Rekadwad et al., 2024). This chapter aims to offer a wide overview of software-solutions, and services for identification of microorganisms.

Basic Local Alignment Search Tool (BLAST)

The Basic Local Alignment Search Tool (BLAST) is a program that searches for regions of similarity between different biological sequences. It is the program's responsibility to compare sequences of nucleotides or proteins to sequence databases and then determine the statistical significance of the results (NCBI-BLAST, 2024). BLAST was created as a speedier, yet arguably equally sensitive, alternative to FASTA for doing similarity searches between DNA and protein sequences. Unlike the dynamic programming algorithm, which has an underlying guarantee of an optimal solution, both of these approaches rely on a heuristic (tried-and-true) approach that virtually always finds related sequences in a database search. FASTA connects query and database sequences into an alignment by identifying brief common patterns. BLAST is comparable to FASTA, except it searches for rarer, more important patterns in protein and nucleic acid sequences, which speeds up the process even more. The fact that BLAST is accessible over the Internet via a sizable server at the National Center for Biotechnology Information (NCBI) and numerous other websites has led to its immense popularity. With its evolution, the BLAST algorithm now offers molecular biologists an array of extremely potent search tools that may be used for free on a variety of computer systems. This page aims to serve as a "user's guide" to the fundamental ideas of BLAST (Altschul et al., 1990; Mount, 2007). A database of sequences is compared with nucleotide or protein sequences using the program, which then determines the statistical significance of the matches. The BLAST Quick Start mini-course (www.ncbi.nlm.nih.gov/Class/minicourses) serves as the basis for this chapter's practical application of several BLAST programs, which begins with an introduction to BLAST. While pertinent theory is included where it influences the selection of the BLAST software, parameters, and database, the focus in each example is on practical, step-by-step operations (Wheeler and Bhagwat, 2007; Camacho et al., 2009; Saeed and Usman, 2019; Alachiotis et al., 2022).

Pictorial representation of steps for analysis, *e.g., nuleotide to nucleotide blast.*

Step-I: Visit to https://blast.ncbi.nlm.nih.gov/Blast.cgi

1. Using the NCBI Homepage

   - Go to the NCBI homepage: https://www.ncbi.nlm.nih.gov/
   - Locate the BLAST link. You will find it in the popular resources section or in the top navigation bar.
   - Click on the BLAST link. This will take you to the main BLAST page with different BLAST options.

2. Direct Link

   - Access the main BLAST page directly using this link: https://blast.ncbi.nlm.nih.gov/Blast.cgi
   - Choosing the Right BLAST Page

The main BLAST page offers several types of BLAST searches:

- *Nucleotide BLAST*: For comparing a nucleotide query sequence against a nucleotide sequence database.
- *Protein BLAST*: For comparing a protein query sequence against a protein sequence database.
- *Translated BLAST searches (blastx or tblastn)*: For comparing translated nucleotide queries against protein databases or protein queries against translated nucleotide databases.
- *BLAST Genomes*: For searching specifically against genomes

Step-II: *Click nucleotide to nucleotide blast. It will open following page.*

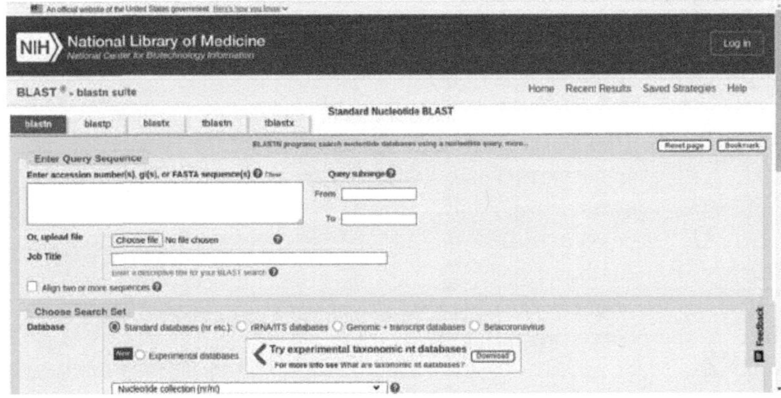

Step-III: *Paste nucleotide sequence or enter accession number of available sequence in the Enter Query Sequence in blastn suite and leave default parameters.*

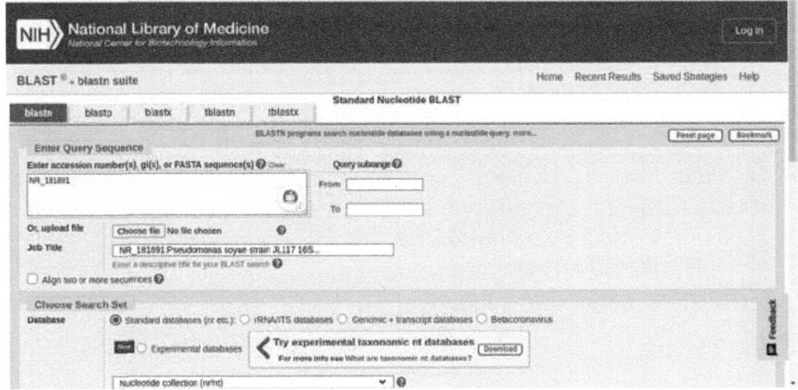

Representative Nucleotide sequence in information available on NCBI-Nucleotide (https://www.ncbi.nlm.nih.gov/nuccore):

```
LOCUS       NR_181891    1532 bp    rRNA    linear    BCT 29-NOV-2022
DEFINITION  Pseudomonas soyae strain JL117 16S ribosomal RNA, partial
            sequence.
ACCESSION   NR_181891
VERSION     NR_181891.1
DBLINK      Project: 33175
            BioProject: PRJNA33175
KEYWORDS    RefSeq.
SOURCE      Pseudomonas soyae
ORGANISM    Pseudomonas soyae
            Bacteria; Pseudomonadota; Gammaproteobacteria; Pseudomonadales;
            Pseudomonadaceae; Pseudomonas.
REFERENCE   1 (bases 1 to 1532)
  AUTHORS   Lemaire,J., Seaton,S., Inderbitzin,P. and Trujillo,M.E.
  TITLE     Pseudomonas zeiradicis from corn and Pseudomonas soyae from
            soybean, two new Pseudomonas species and endophytes from
            agricultural crops
  JOURNAL   bioRxiv (2021)
REFERENCE   2 (bases 1 to 1532)
  CONSRTM   NCBI RefSeq Targeted Loci Project
  TITLE     Direct Submission
  JOURNAL   Submitted (07-OCT-2022) National Center for Biotechnology
            Information, Nih, Bethesda, MD 20894, USA
REFERENCE   3 (bases 1 to 1532)
  AUTHORS   Seaton,S., Lemaire,J. and Inderbitzin,P.
  TITLE     Direct Submission
  JOURNAL   Submitted (28-OCT-2021) R and D, Indigo Ag, 500 Rutherford Ave,
            Boston, MA 02129, USA
COMMENT     REVIEWED REFSEQ: This record has been curated by
            NCBI staff.
            The reference sequence is identical to OL306321.1:1-1532.

            ##Assembly-Data-START##
            Assembly Method :: Megahit v. 1.1.2
            Sequencing Technology :: Illumina
            ##Assembly-Data-END##
FEATURES    Location/Qualifiers
     source  1..1532
             /organism="Pseudomonas soyae"
             /mol_type="rRNA"
```

```
                    /strain="JL117"
                    /host="Glycine max"
                    /type_material="type strain of Pseudomonas soyae"
                    /db_xref="taxon:2890313"
                    /country="USA: Indiana"
     rRNA           <1..>1532
                    /product="16S ribosomal RNA"
ORIGIN
    1   tgaagagttt gatcatggct cagattgaac gctggcggca ggcctaacac atgcaagtcg
   61   agcggatgac aggagcttgc tcctgaattc agcggcggac gggtgagtaa tgcctaggaa
  121   tctgcctggt agtgggggac aacgtttcga aaggaacgct aataccgcat acgtcctacg
  181   ggagaaagca ggggaccttc gggccttgcg ctatcagatg agcctaggtc ggattagcta
  241   gttggtgagg taatggctca ccaaggcgac gatccgtaac tggtctgaga ggatgatcag
  301   tcacactgga actgagacac ggtccagact cctacgggag gcagcagtgg ggaatattgg
  361   acaatgggcg aaagcctgat ccagccatgc cgcgtgtgtg aagaaggtct tcggattgta
  421   aagcacttta agttgggagg aagggttgta gattaatact ctgcaatttt gacgttaccg
  481   acagaataag caccggctaa ctctgtgcca gcagccgcgg taatacagag ggtgcaagcg
  541   ttaatcggaa ttactgggcg taaagcgcgc gtaggtggtt cgttaagttg gatgtgaaat
  601   ccccgggctc aacctgggaa ctgcattcaa aactgtcgag ctagagtatg gtagagggtg
  661   gtggaatttc ctgtgtagcg gtgaaatgcg tagatatagg aaggaacacc agtggcgaag
  721   gcgaccacct ggactgatac tgacactgag gtgcgaaagc gtggggagca aacaggatta
  781   gataccctgg tagtccacgc cgtaaacgat gtcaactagc cgttgggagc cttgagctct
  841   tagtggcgca gctaacgcat taagttgacc gcctggggag tacggccgca aggttaaaac
  901   tcaaatgaat tgacgggggc ccgcacaagc ggtggagcat gtggtttaat tcgaagcaac
  961   gcgaagaacc ttaccaggcc ttgacatcca atgaactttc cagagatgga ttggtgcctt
 1021   cgggagcatt gagacaggtg ctgcatggct gtcgtcagct cgtgtcgtga gatgttgggt
 1081   taagtcccgt aacgagcgca acccttgtcc ttagttacca gcacgttatg gtgggcactc
 1141   taaggagact gccggtgaca aaccggagga aggtggggat gacgtcaagt catcatggcc
 1201   cttacggcct gggctacaca cgtgctacaa tggtcggtac aaagggttgc caagccgcga
 1261   ggtggagcta atcccataaa accgatcgta gtccggatcg cagtctgcaa ctcgactgcg
 1321   tgaagtcgga atcgctagta atcgtgaatc agaatgtcac ggtgaatacg ttcccgggcc
 1381   ttgtacacac cgcccgtcac accatgggag tgggttgcac cagaagtagc tagtctaacc
 1441   ttcgggagga cggttaccac ggtgtgattc atgactgggg tgaagtcgta acaaggtagc
 1501   cgtaggggaa cctgcggctg gatcacctcc tt
        //
```

The Above information needs to be obtained in FASTA format for NCBI-Blast analysis.

Representative Nucleotide sequence in FASTA format for NCBI-BLAST (Note: own research data):

>JN392969.1 Brevibacillus borstelensis strain EF_NAK1-7 16S ribosomal RNA gene, partial sequence
GAAGGGCGGCAGCTATACATGCAAGTCGAGCGAGTCCCTTCGGGGGC
TAGCGGCGGACGGGTGAGTAACA
CGTAGGCAACCTGCCCGTAAGCTCGGGATAACATGGGGAAACTCATG
CTAATACCGGATAGGGTCTTCTC
TCGCATGAGAGGAGACGGAAAGGTGGCGCAAGCTACCACTTACGGA
TGGGCCTGCGGCGCATTAGCTAGT
TGGTGGGGTAACGGCCTACCAAGGCGACGATGCGTAGCCGACCTGAG
AGGGTGACCGGCCACACTGGGAC
TGAGACACGGCCCAGACTCCTACGGGAGGCAGCAGTAGGGAATTTTC
CACAATGGACGAAAGTCTGATGG
AGCAACGCCGCGTGAACGATGAAGGTCTTCGGATTGTAAAGTTCTGT
TGTCAGAGACGAACAAGTACCGT
TCGAACAGGGCGGTACCTTGACGGTACCTGACGAGAAAGCCACGGC
TAACTACGTGCCAGCAGCCGCGGT
AATACGTAGGTGGCAAGCGTTGTCCGGAATTATTGGGCGTAAAGCGC
GCGCAGGCGGCTATGTAAGTCTG
GTGTTAAAGCCCGGGGCTCAACCCCGGTTCGCATCGGAAACTGTGTA
GCTTGAGTGCAGAAGAGGAAAGC
GGTATTCCACGTGTAGCGGTGAAATGCGTAGAGATGTGGAGGAACAC
CAGTGGCGAAGGCGGCTTTCTGG
TCTGTAACTGACGCTGAGGCGCGAAAGCGTGGGGAGCAAACAGGAT
TAGATACCCTGGGTAGTCCACGCC
GTAACGATGAGTGCTAGTGTGGGGGGTTTCATACCCTCAGTGCCGCA
GCTAACGCAATAAGCCACCTCCC
CGCCTGGGGGGAGG

>JN392970.1 Brevibacillus borstelensis strain EF_TYK1-4 16S ribosomal RNA gene, partial sequence
GGGGCCTTGGGGGCGGCTATACATGCAAGTCGAGCGAGTCCCTTCG
GGGGCTAGCGGCGGACGGGTGAGT
AACACGTAGGCAACCTGCCCGTAAGCTCGGGATAACATGGGGAAA
CTCATGCTAATACCGGATAGGGTCT
TCTCTCGCATGAGAGGAGACGGAAAGGTGGCGCAAGCTACCACTT
ACGGATGGGCCTGCGGCGCATTAGC
TAGTTGGTGGGGTAACGGCCTACCAAGGCGACGATGCGTAGCCGA
CCTGAGAGGGTGACCGGCCACACTG
GGACTGAGACACGGCCCAGACTCCTACGGGAGGCAGCAGTAGGG
AATTTTCCACAATGGACGAAAGTCTG
ATGGAGCAACGCCGCGTGAACGATGAAGGTCTTCGGATTGTAAAG
TTCTGTTGTCAGAGACGAACAAGTA

CCGTTCGAACAGGGCGGTACCTTGACGGTACCTGACGAGAAAGCC
ACGGCTAACTACGTGCCAGCAGCCG
CGGTAATACGTAGGTGGCAAGCGTTGTCCGGAATTATTGGGCGTAA
AGCGCGCGCAGGCGGCTATGTAAG
TCTGGTGTTAAAGCCCGGGGCTCAACCCCGGTTCGCATCGGAAACT
GTGTAGCTTGAGTGCAGAAGAGGA
AAGCGGTATTCCACGTGTAGCGGTGAAATGCGTAGAGATGTGGAG
GAACACCAGTGGCGAAGGCGGCTTT
CTGGTCTGTAACTGACGCTGAGGCGCGAAAGCGTGGGGAGCAAA
CAGGATTAGATACCCTGGTAGTCCAC
GCCCGTAACGATGAGTGCTAGTGTTTGGGGGGTTTCAATACCCTCA
GTGCCGCAGCTAACGCATAGCACT
CCGCCCTGGGGGAGTACGCTCGCAAA

>*JN392971.1 Bacillus paralicheniformis strain EF_TYK1-5 16S ribosomal RNA gene, partial sequence*
GCATGGCGGCAGCTAATACATGCAGTCGAGCGGACCGACGGGAGCT
TGCTCCCTTAGGTCAGCGGCGGAC
GGGTGAGTAACACGTGGGTAACCTGCCTGTAAGACTGGGATAACTC
CGGGAAACCGGGGCTAATACCGGA
TGCTTGATTGAACCGCATGGTTCAATCATAAAAGGTGGCTTTTAGCT
ACCACTTACAGATGGACCCGCGG
CGCATTAGCTAGTTGGTGAGGTAACGGCTCACCAAGGCGACGATGC
GTAGCCGACCTGAGAGGGTGATCG
GCCACACTGGGACTGAGACACGGCCCAGACTCCTACGGGAGGCAG
CAGTAGGGAATCTTCCGCAATGGAC
GAAAGTCTGACGGAGCAACGCCGCGTGAGTGATGAAGGTTTTCGG
ATCGTAAAACTCTGTTGTTAGGGAA
GAACAAGTACCGTTCGAATAGGGCGGCACCTTGACGGTACCTAACC
AGAAAGCCACGGCTAACTACGTGC
CAGCAGCCGCGGTAATACGTAGGTGGCAAGCGTTGTCCGGAATTAT
TGGGCGTAAAGCGCGCGCAGGCGG
TTTCTTAAGTCTGATGTGAAAGCCCCCGGCTCAACCGGGGAGGGTC
ATTGGAAACTGGGGGAACTTGAGT
GCAGAAGAGGAGAGTGGAATTCCACGTGTAGCGGTGAAATGCGTA
GAGATGTGGAGGAACACCAGTGGCG
AAAGGCGACTCTCTGGTCTGTAACTGACGCTGAGGCGCGAAAGCG
TGGGGAGCGAACAGGATTAGATACC
CTGGTAGTCCACGCCGTAACGATGAGTGCTAAGTGTAGAGGGTTCC
GGCCCTTAGTGCTGCAGCAAACGC
ATTAAGCACCTCCCGCCCTGAGA

>KC120909.1 *Bacillus altitudinis* strain W10 16S ribosomal RNA gene, partial sequence
TGATGTGAAAGCCCCCAGCTCAACCGGGGAAGGGTCATTGGAAAC
TGGGAAACTTGAGTGCAGAAGAGGA
GAGTGGAATTCCACGTGTAGCGGTGAAATGCGTAGAGATGTGGAG
GAACACCAGTGGCGAAGGCGACTCT
CTGGTCTGTAACTGACGCTGAGGAGCGAAAGCGTGGGGAGCGAA
CAGGATTAGATACCCTGGTAGTCCAC
GCCGTAAACGATGAGTGCTAAGTGTTAGGGGGTTTCCGCCCCTTAG
TGCTGCAGCTAACGCATTAAGCAC
TCCGCCTGGGGAGTACGGTCGCAAGACTGAAACTCAAAGGAATTG
ACGGGGGCCCGCACAAGCGGTGGAG
CATGTGGTTTAATTCGAAGCAACGCGAAGAACCTTACCAGGTCTTG
ACATCCTCTGACAACCCTAGAGAT
AGGGCTTTCCCTTCGGGGACAGAGTGACAGGTGGTGCATGGTTGTC
GTCAGCTCGTGTCGTGAGATGTTG
GGTTAAGTCCCGCAACGAGCGCAACCCTTGATCTTAGTTGCCAGCA
TTCAGTTGGGCACTCTAAGGTGAC
TGCCGGTGACAAACCGGAGGAAGGTGGGGATGACGTCAAATCATC
ATGCCCCTTATGACCTGGGCTACAC
ACGTGCTACAATGGACAGAACAAAGGGCTGCGAGACCGCAAGGTT
TAGCCAATCCCACAAATCTGTTCTC
AGTTCGGATCGCAGTCTGCAACTCGACTGCGTGAAGCTGGAATCGC
TAGTAATCGCGGATCAGCATGCCG
CGGTGAATACGTTCCCGGGCCTTGTACACACCGCCCGTCACACCAC
GAGAGTTTGCAACACCCGAAGTCG
GTGAGGTAACCTTTATGGAGCCAGCCGCCGAA

>KC120910.1 *Brevibacillus borstelensis* strain NAK1-9 16S ribosomal RNA gene, partial sequence
GAAGTAGGATGGACGTCAGCAGCGCGTGTCGGATTATTGGGCGTA
AAGCGCGCGCAGGCGGCTATGTAAG
TCTGGTGTTAAAGCCCGGGGCTCAACCCCGGTTCGCATCGGAAAC
TGTGTAGCTTGAGTGCAGAAGAGGA
AAGCGGTATTCCACGTGTAGCGGTGAAATGCGTAGAGATGTGGAG
GAACACCAGTGGCGAAGGCGGCTTT
CTGGTCTGTAACTGACGCTGAGGCGCGAAAGCGTGGGGAGCAAA
CAGGATTAGATACCCTGGTAGTCCAC
GCCGTAAACGATGAGTGCTAGGTGTTGGGGGTTTCAATACCCTCA
GTGCCGCAGCTAACGCAATAAGCAC
TCCGCCTGGGGAGTACGCTCGCAAGAGTGAAACTCAAAGGAATTG
ACGGGGGCCCGCACAAGCGGTGGAG
CATGTGGTTTAATTCGAAGCAACGCGAAGAACCTTACCAGGTCTTG
ACATCCCGCTGACCGTCCTAGAGA

TAGGGCTTCCCTTCGGGGCAGCGGTGACAGGTGGTGCATGGTTGTC
GTCAGCTCGTGTCGTGAGATGTTG
GGTTAAGTCCCGCAACGAGCGCAACCCTTATCTTTAGTTGCCAGCA
TTCAGTTGGGCACTCTAGAGAGAC
TGCCGTCGACAAGACGGAGGAAGGCGGGGATGACGTCAAATCATC
ATGCCCCTTATGACCTGGGCTACAC
ACGTGCTACAATGGCTGGTACAACGGGAAGCTAGCTCGCGAGAGT
ATGCCAATCTCTTAAAACCAGTCTC
AGTTCGGATTGCAGGCTGCAACTCGCCTGCATGAAGTCGGAATCG
CTAGTAATCGCGGATCAGCATGCCG
CGTGAATACGTTCCCGGGCTTCTC

>*KC120911.1 Brevibacterium aurantiacum strain W7 16S ribosomal RNA gene, partial sequence*
GCTCACCAAGACGACGACGGGTAGCCGGCCTGAGAGGGCGACCG
GCCACACTGGGACTGAGACACGGCCC
AGACTCCTACGGGAGGCAGCAGTGGGGAATATTGCACAATGGGGG
AAACCCTGATGCAGCGACGCAGCGT
GCGGGATGACGGCCTTCGGGTTGTAAACCGCTTTCAGCAGGGAAG
AAGCCCCTTGGGGTGACGGTACCTG
CAGAAGAAGTACCGGCTAACTACGTGCCAGCAGCCGCGGTAATAC
GTAGGGTACAAGCGTTGTCCGGAAT
TATTGGGCGTAAAGAGCTCGTAGGTGGTTGGTCACGTCTGCTGTGG
AAACGCAACGCTTAACGTTGCGCG
TGCAGTGGGTACGGGCTGACTAGAGTGCAATAGGGGAGTCTGGAA
TTCCTGGTGTAGCGGTGAAATGCGC
AGATATCAGGAGGAACACCGGTGGCGAAGGCGGGACTCTGGGCTG
TAACTGACACTGAGGAGCGAAAGCA
TGGGGAGCGAACAGGATTAGATACCCTGGTAGTCCATGCCGTAAAC
GTTGGGCACTAGGTGTGGGGGACA
TTCCACGTTCTCCGCGCCGTAGCTAACGCATTAAGTGCCCCGCCTG
GGGAGTACGGTCGCAAGGCTAAAA
CTCAAAGGAATTGACGGGGGCCCGCACAAGCGGCGGAGCATGCG
GATTAATTCGATGCAACGCGAAGAAC
CTTACCAAGGCTTGACATACACCAGACCGTTCTGGAAACAGTTCCT
CCCCTTTGGGGTTGGTGTACAGGT
GGTGCATGGTTGTCGTCAGCTCGTGTCGTGAGATGTTGGGTTAAGT
CCCGCAACGAGCGCAACCCTCGTT
CTATGTTGCCAGCACGTAATGGTGGGAACTCATAGGAGACTGCCGG
GGTCAACTCGGAGGAAGGTGGGGA
TGACGTCAAATCATCATGCCCTTTATGTCTTGGGCTTCACGCATGCT
ACAATGGCTGGTACAGAGAGATG
CGAGA

Step-IV: Your query sequence will be analyzed by BLAST

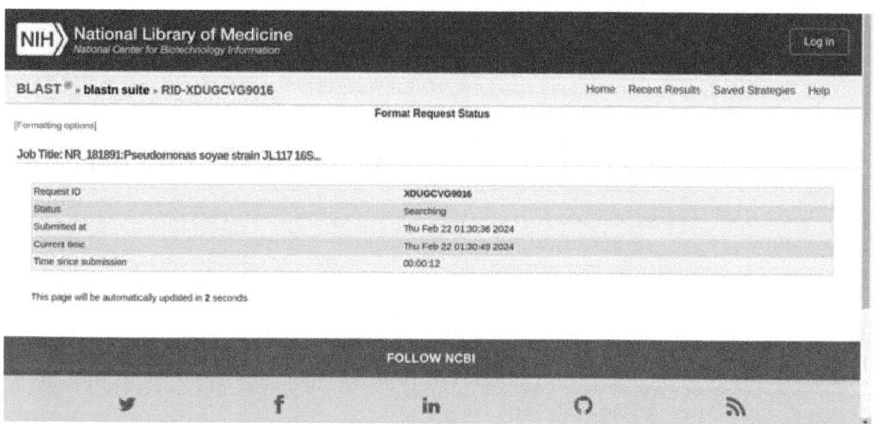

Step-V: Representative output of BLAST analysis

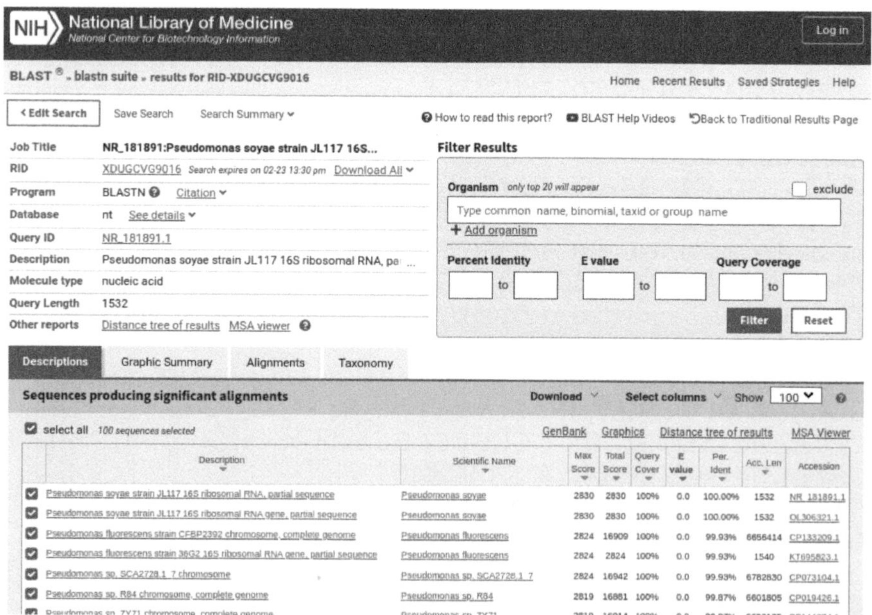

Step-VI: Analyse output of BLAST analysis (Source: https://blast.ncbi.nlm.nih. gov/Blast.cgi)

Step-VII: Check the alignment statistics for top at least top 10 strains on NCBI-BLAST server against sequence of interest used for BLAST analysis. It shows score, identities of bases matched, gaps in the sequence.

Pseudomonas soyae strain JL117 16S ribosomal RNA, partial sequence NR_181891.1 1532 1

GenBank Graphics Next Match Previous Match
Alignment statistics for match #1

Score	Expect	Identities	Gaps	Strand
2830 bits(1532)	0.0	1532/1532(100%)	0/1532(0%)	Plus/Plus

Query 1   *TGAAGAGTTTGATCATGGCTCAGATTGAACGCTGGCGGCA GGCCTAACACATGCAAGTCG 60*

    |||||||||||||||||||||||||||||||||||||||||||||||||||||||||||||

Sbjct 1   *TGAAGAGTTTGATCATGGCTCAGATTGAACGCTGGCGGCA GGCCTAACACATGCAAGTCG 60*

Query 61   *AGCGGATGACAGGAGCTTGCTCCTGAATTCAGCGGCGGA CGGGTGAGTAATGCCTAGGAA 120*

    |||||||||||||||||||||||||||||||||||||||||||||||||||||||||||||

Sbjct 61   *AGCGGATGACAGGAGCTTGCTCCTGAATTCAGCGGCGGA CGGGTGAGTAATGCCTAGGAA 120*

Query 121   *TCTGCCTGGTAGTGGGGGACAACGTTTCGAAAGGAACGC TAATACCGCATACGTCCTACG 180*

    |||||||||||||||||||||||||||||||||||||||||||||||||||||||||||||

Sbjct 121   *TCTGCCTGGTAGTGGGGGACAACGTTTCGAAAGGAACGC TAATACCGCATACGTCCTACG 180*

Query 181   *GGAGAAAGCAGGGGACCTTCGGGCCTTGCGCTATCAGAT GAGCCTAGGTCGGATTAGCTA 240*

    |||||||||||||||||||||||||||||||||||||||||||||||||||||||||||||

Sbjct 181   *GGAGAAAGCAGGGGACCTTCGGGCCTTGCGCTATCAGAT GAGCCTAGGTCGGATTAGCTA 240*

```
Query 241   GTTGGTGAGGTAATGGCTCACCAAGGCGACGATCCGTAAC
            TGGTCTGAGAGGATGATCAG 300
            ||||||||||||||||||||||||||||||||||||||||
            ||||||
Sbjct 241   GTTGGTGAGGTAATGGCTCACCAAGGCGACGATCCGTAA
            CTGGTCTGAGAGGATGATCAG 300

Query 301   TCACACTGGAACTGAGACACGGTCCAGACTCCTACGGGA
            GGCAGCAGTGGGGAATATTGG 360
            ||||||||||||||||||||||||||||||||||||||||
            ||||||
Sbjct 301   TCACACTGGAACTGAGACACGGTCCAGACTCCTACGGGA
            GGCAGCAGTGGGGAATATTGG 360

Query 361   ACAATGGGCGAAAGCCTGATCCAGCCATGCCGCGTGTGT
            GAAGAAGGTCTTCGGATTGTA 420
            ||||||||||||||||||||||||||||||||||||||||
            ||||||
Sbjct 361   ACAATGGGCGAAAGCCTGATCCAGCCATGCCGCGTGTGT
            GAAGAAGGTCTTCGGATTGTA 420

Query 421   AAGCACTTTAAGTTGGGAGGAAGGGTTGTAGATTAATACT
            CTGCAATTTTGACGTTACCG 480
            ||||||||||||||||||||||||||||||||||||||||
            ||||||
Sbjct 421   AAGCACTTTAAGTTGGGAGGAAGGGTTGTAGATTAATACT
            CTGCAATTTTGACGTTACCG 480

Query 481   ACAGAATAAGCACCGGCTAACTCTGTGCCAGCAGCCGCG
            GTAATACAGAGGGTGCAAGCG 540
            ||||||||||||||||||||||||||||||||||||||||
            ||||||
Sbjct 481   ACAGAATAAGCACCGGCTAACTCTGTGCCAGCAGCCGCG
            GTAATACAGAGGGTGCAAGCG 540

Query 541   TTAATCGGAATTACTGGGCGTAAAGCGCGCGTAGGTGGTT
            CGTTAAGTTGGATGTGAAAT 600
            ||||||||||||||||||||||||||||||||||||||||
            ||||||
Sbjct 541   TTAATCGGAATTACTGGGCGTAAAGCGCGCGTAGGTGGTT
            CGTTAAGTTGGATGTGAAAT 600
```

```
Query 601  CCCCGGGCTCAACCTGGGAACTGCATTCAAAACTGTCGAG
           CTAGAGTATGGTAGAGGGTG 660
           ||||||||||||||||||||||||||||||||||||||||
           ||||||
Sbjct 601  CCCCGGGCTCAACCTGGGAACTGCATTCAAAACTGTCGA
           GCTAGAGTATGGTAGAGGGTG 660

Query 661  GTGGAATTTCCTGTGTAGCGGTGAAATGCGTAGATATAGG
           AAGGAACACCAGTGGCGAAG 720
           ||||||||||||||||||||||||||||||||||||||||
           ||||||
Sbjct 661  GTGGAATTTCCTGTGTAGCGGTGAAATGCGTAGATATAGG
           AAGGAACACCAGTGGCGAAG 720

Query 721  GCGACCACCTGGACTGATACTGACACTGAGGTGCGAAAG
           CGTGGGGAGCAAACAGGATTA 780
           ||||||||||||||||||||||||||||||||||||||||
           ||||||
Sbjct 721  GCGACCACCTGGACTGATACTGACACTGAGGTGCGAAAG
           CGTGGGGAGCAAACAGGATTA 780

Query 781  GATACCCTGGTAGTCCACGCCGTAAACGATGTCAACTAGC
           CGTTGGGAGCCTTGAGCTCT 840
           ||||||||||||||||||||||||||||||||||||||||
           ||||||
Sbjct 781  GATACCCTGGTAGTCCACGCCGTAAACGATGTCAACTAGC
           CGTTGGGAGCCTTGAGCTCT 840

Query 841  TAGTGGCGCAGCTAACGCATTAAGTTGACCGCCTGGGGAG
           TACGGCCGCAAGGTTAAAAC 900
           ||||||||||||||||||||||||||||||||||||||||
           ||||||
Sbjct 841  TAGTGGCGCAGCTAACGCATTAAGTTGACCGCCTGGGGAG
           TACGGCCGCAAGGTTAAAAC 900

Query 901  TCAAATGAATTGACGGGGGCCCGCACAAGCGGTGGAGCA
           TGTGGTTTAATTCGAAGCAAC 960
           ||||||||||||||||||||||||||||||||||||||||
           ||||||
Sbjct 901  TCAAATGAATTGACGGGGGCCCGCACAAGCGGTGGAGCA
           TGTGGTTTAATTCGAAGCAAC 960
```

Query 961   GCGAAGAACCTTACCAGGCCTTGACATCCAATGAACTTTC
CAGAGATGGATTGGTGCCTT 1020
            ||||||||||||||||||||||||||||||||||||||||
            ||||||
Sbjct 961   GCGAAGAACCTTACCAGGCCTTGACATCCAATGAACTTTC
CAGAGATGGATTGGTGCCTT 1020

Query 1021 CGGGAGCATTGAGACAGGTGCTGCATGGCTGTCGTCAGCT
CGTGTCGTGAGATGTTGGGT 1080
            ||||||||||||||||||||||||||||||||||||||||
            ||||||
Sbjct 1021 CGGGAGCATTGAGACAGGTGCTGCATGGCTGTCGTCAGCT
CGTGTCGTGAGATGTTGGGT 1080

Query 1081 TAAGTCCCGTAACGAGCGCAACCCTTGTCCTTAGTTACCA
GCACGTTATGGTGGGCACTC 1140
            ||||||||||||||||||||||||||||||||||||||||
            ||||||
Sbjct 1081 TAAGTCCCGTAACGAGCGCAACCCTTGTCCTTAGTTACCA
GCACGTTATGGTGGGCACTC 1140

Query 1141 TAAGGAGACTGCCGGTGACAAACCGGAGGAAGGTGGGG
ATGACGTCAAGTCATCATGGCC 1200
            ||||||||||||||||||||||||||||||||||||||||
            ||||||
Sbjct 1141 TAAGGAGACTGCCGGTGACAAACCGGAGGAAGGTGGGG
ATGACGTCAAGTCATCATGGCC 1200

Query 1201 CTTACGGCCTGGGCTACACACGTGCTACAATGGTCGGTAC
AAAGGGTTGCCAAGCCGCGA 1260
            ||||||||||||||||||||||||||||||||||||||||
            ||||||
Sbjct 1201 CTTACGGCCTGGGCTACACACGTGCTACAATGGTCGGTAC
AAAGGGTTGCCAAGCCGCGA 1260

Query 1261 GGTGGAGCTAATCCCATAAAACCGATCGTAGTCCGGATCG
CAGTCTGCAACTCGACTGCG 1320
            ||||||||||||||||||||||||||||||||||||||||
            ||||||
Sbjct 1261 GGTGGAGCTAATCCCATAAAACCGATCGTAGTCCGGATCG
CAGTCTGCAACTCGACTGCG 1320

```
Query 1321  TGAAGTCGGAATCGCTAGTAATCGTGAATCAGAATGTCAC
            GGTGAATACGTTCCCGGGCC 1380
            ||||||||||||||||||||||||||||||||||||||||
            ||||||
Sbjct 1321  TGAAGTCGGAATCGCTAGTAATCGTGAATCAGAATGTCAC
            GGTGAATACGTTCCCGGGCC 1380

Query 1381  TTGTACACACCGCCCGTCACACCATGGGAGTGGGTTGCAC
            CAGAAGTAGCTAGTCTAACC 1440
            ||||||||||||||||||||||||||||||||||||||||
            ||||||
Sbjct 1381  TTGTACACACCGCCCGTCACACCATGGGAGTGGGTTGCAC
            CAGAAGTAGCTAGTCTAACC 1440

Query 1441  TTCGGGAGGACGGTTACCACGGTGTGATTCATGACTGGGG
            TGAAGTCGTAACAAGGTAGC 1500
            ||||||||||||||||||||||||||||||||||||||||
            ||||||
Sbjct 1441  TTCGGGAGGACGGTTACCACGGTGTGATTCATGACTGGGG
            TGAAGTCGTAACAAGGTAGC 1500

Query 1501  CGTAGGGGAACCTGCGGCTGGATCACCTCCTT 1532
            ||||||||||||||||||||||||||||||||
Sbjct 1501  CGTAGGGGAACCTGCGGCTGGATCACCTCCTT 1532
```

Download GenBank Graphics Next Previous Descriptions

Pseudomonas soyae strain JL117 16S ribosomal RNA gene, partial sequence
OL306321.1 1532 1

GenBank Graphics Next Match Previous Match
Alignment statistics for match #1

Score	Expect	Identities	Gaps	Strand
2830 bits(1532)	0.0	1532/1532(100%)	0/1532(0%)	Plus/Plus

```
Query 1     TGAAGAGTTTGATCATGGCTCAGATTGAACGCTGGCGGCA
            GGCCTAACACATGCAAGTCG 60
            ||||||||||||||||||||||||||||||||||||||||
            ||||||
Sbjct 1     TGAAGAGTTTGATCATGGCTCAGATTGAACGCTGGCGGCA
            GGCCTAACACATGCAAGTCG 60
```

Query 61   AGCGGATGACAGGAGCTTGCTCCTGAATTCAGCGGCGGA
             CGGGTGAGTAATGCCTAGGAA 120
             |||||||||||||||||||||||||||||||||||||||||||||||||
             ||||||

Sbjct 61   AGCGGATGACAGGAGCTTGCTCCTGAATTCAGCGGCGGA
             CGGGTGAGTAATGCCTAGGAA 120

Query 121   TCTGCCTGGTAGTGGGGGACAACGTTTCGAAAGGAACGC
             TAATACCGCATACGTCCTACG 180
             |||||||||||||||||||||||||||||||||||||||||||||||||
             ||||||

Sbjct 121   TCTGCCTGGTAGTGGGGGACAACGTTTCGAAAGGAACGC
             TAATACCGCATACGTCCTACG 180

Query 181   GGAGAAAGCAGGGGACCTTCGGGCCTTGCGCTATCAGAT
             GAGCCTAGGTCGGATTAGCTA 240
             |||||||||||||||||||||||||||||||||||||||||||||||||
             ||||||

Sbjct 181   GGAGAAAGCAGGGGACCTTCGGGCCTTGCGCTATCAGAT
             GAGCCTAGGTCGGATTAGCTA 240

Query 241   GTTGGTGAGGTAATGGCTCACCAAGGCGACGATCCGTAAC
             TGGTCTGAGAGGATGATCAG 300
             |||||||||||||||||||||||||||||||||||||||||||||||||
             ||||||

Sbjct 241   GTTGGTGAGGTAATGGCTCACCAAGGCGACGATCCGTAAC
             TGGTCTGAGAGGATGATCAG 300

Query 301   TCACACTGGAACTGAGACACGGTCCAGACTCCTACGGGA
             GGCAGCAGTGGGGAATATTGG 360
             |||||||||||||||||||||||||||||||||||||||||||||||||
             ||||||

Sbjct 301   TCACACTGGAACTGAGACACGGTCCAGACTCCTACGGGA
             GGCAGCAGTGGGGAATATTGG 360

Query 361   ACAATGGGCGAAAGCCTGATCCAGCCATGCCGCGTGTGTG
             AAGAAGGTCTTCGGATTGTA 420
             |||||||||||||||||||||||||||||||||||||||||||||||||
             ||||||

Sbjct 361   ACAATGGGCGAAAGCCTGATCCAGCCATGCCGCGTGTGTG
             AAGAAGGTCTTCGGATTGTA 420

Query 421   AAGCACTTTAAGTTGGGAGGAAGGGTTGTAGATTAATACT
            CTGCAATTTTGACGTTACCG 480
            ||||||||||||||||||||||||||||||||||||||||
            ||||||
Sbjct 421   AAGCACTTTAAGTTGGGAGGAAGGGTTGTAGATTAATACT
            CTGCAATTTTGACGTTACCG 480

Query 481   ACAGAATAAGCACCGGCTAACTCTGTGCCAGCAGCCGCG
            GTAATACAGAGGGTGCAAGCG 540
            ||||||||||||||||||||||||||||||||||||||||
            ||||||
Sbjct 481   ACAGAATAAGCACCGGCTAACTCTGTGCCAGCAGCCGCG
            GTAATACAGAGGGTGCAAGCG 540

Query 541   TTAATCGGAATTACTGGGCGTAAAGCGCGCGTAGGTGGTT
            CGTTAAGTTGGATGTGAAAT 600
            ||||||||||||||||||||||||||||||||||||||||
            ||||||
Sbjct 541   TTAATCGGAATTACTGGGCGTAAAGCGCGCGTAGGTGGTT
            CGTTAAGTTGGATGTGAAAT 600

Query 601   CCCCGGGCTCAACCTGGGAACTGCATTCAAAACTGTCGA
            GCTAGAGTATGGTAGAGGGTG 660
            ||||||||||||||||||||||||||||||||||||||||
            ||||||
Sbjct 601   CCCCGGGCTCAACCTGGGAACTGCATTCAAAACTGTCGAG
            CTAGAGTATGGTAGAGGGTG 660

Query 661   GTGGAATTTCCTGTGTAGCGGTGAAATGCGTAGATATAGG
            AAGGAACACCAGTGGCGAAG 720
            ||||||||||||||||||||||||||||||||||||||||
            ||||||
Sbjct 661   GTGGAATTTCCTGTGTAGCGGTGAAATGCGTAGATATAGG
            AAGGAACACCAGTGGCGAAG 720

Query 721   GCGACCACCTGGACTGATACTGACACTGAGGTGCGAAAG
            CGTGGGGAGCAAACAGGATTA 780
            ||||||||||||||||||||||||||||||||||||||||
            ||||||
Sbjct 721   GCGACCACCTGGACTGATACTGACACTGAGGTGCGAAAG
            CGTGGGGAGCAAACAGGATTA 780

Query 781   GATACCCTGGTAGTCCACGCCGTAAACGATGTCAACTAGC
CGTTGGGAGCCTTGAGCTCT 840
            ||||||||||||||||||||||||||||||||||||||||
||||||
Sbjct 781   GATACCCTGGTAGTCCACGCCGTAAACGATGTCAACTAGC
CGTTGGGAGCCTTGAGCTCT 840

Query 841   TAGTGGCGCAGCTAACGCATTAAGTTGACCGCCTGGGGA
GTACGGCCGCAAGGTTAAAAC 900
            ||||||||||||||||||||||||||||||||||||||||
||||||
Sbjct 841   TAGTGGCGCAGCTAACGCATTAAGTTGACCGCCTGGGGAG
TACGGCCGCAAGGTTAAAAC 900

Query 901   TCAAATGAATTGACGGGGGCCCGCACAAGCGGTGGAGCA
TGTGGTTTAATTCGAAGCAAC 960
            ||||||||||||||||||||||||||||||||||||||||
||||||
Sbjct 901   TCAAATGAATTGACGGGGGCCCGCACAAGCGGTGGAGCA
TGTGGTTTAATTCGAAGCAAC 960

Query 961   GCGAAGAACCTTACCAGGCCTTGACATCCAATGAACTTTC
CAGAGATGGATTGGTGCCTT 1020
            ||||||||||||||||||||||||||||||||||||||||
||||||
Sbjct 961   GCGAAGAACCTTACCAGGCCTTGACATCCAATGAACTTTC
CAGAGATGGATTGGTGCCTT 1020

Query 1021 CGGGAGCATTGAGACAGGTGCTGCATGGCTGTCGTCAGCT
CGTGTCGTGAGATGTTGGGT 1080
            ||||||||||||||||||||||||||||||||||||||||
||||||
Sbjct 1021 CGGGAGCATTGAGACAGGTGCTGCATGGCTGTCGTCAGCT
CGTGTCGTGAGATGTTGGGT 1080

Query 1081 TAAGTCCCGTAACGAGCGCAACCCTTGTCCTTAGTTACCA
GCACGTTATGGTGGGCACTC 1140
            ||||||||||||||||||||||||||||||||||||||||
||||||
Sbjct 1081 TAAGTCCCGTAACGAGCGCAACCCTTGTCCTTAGTTACCA
GCACGTTATGGTGGGCACTC 1140

```
Query 1141  TAAGGAGACTGCCGGTGACAAACCGGAGGAAGGTGGGGA
            TGACGTCAAGTCATCATGGCC 1200
            |||||||||||||||||||||||||||||||||||||||
            ||||||
Sbjct 1141  TAAGGAGACTGCCGGTGACAAACCGGAGGAAGGTGGGGA
            TGACGTCAAGTCATCATGGCC 1200

Query 1201  CTTACGGCCTGGGCTACACACGTGCTACAATGGTCGGTAC
            AAAGGGTTGCCAAGCCGCGA 1260
            |||||||||||||||||||||||||||||||||||||||
            ||||||
Sbjct 1201  CTTACGGCCTGGGCTACACACGTGCTACAATGGTCGGTAC
            AAAGGGTTGCCAAGCCGCGA 1260

Query 1261  GGTGGAGCTAATCCCATAAAACCGATCGTAGTCCGGATCG
            CAGTCTGCAACTCGACTGCG 1320
            |||||||||||||||||||||||||||||||||||||||
            ||||||
Sbjct 1261  GGTGGAGCTAATCCCATAAAACCGATCGTAGTCCGGATCG
            CAGTCTGCAACTCGACTGCG 1320

Query 1321  TGAAGTCGGAATCGCTAGTAATCGTGAATCAGAATGTCAC
            GGTGAATACGTTCCCGGGCC 1380
            |||||||||||||||||||||||||||||||||||||||
            ||||||
Sbjct 1321  TGAAGTCGGAATCGCTAGTAATCGTGAATCAGAATGTCAC
            GGTGAATACGTTCCCGGGCC 1380

Query 1381  TTGTACACACCGCCCGTCACACCATGGGAGTGGGTTGCAC
            CAGAAGTAGCTAGTCTAACC 1440
            |||||||||||||||||||||||||||||||||||||||
            ||||||
Sbjct 1381  TTGTACACACCGCCCGTCACACCATGGGAGTGGGTTGCAC
            CAGAAGTAGCTAGTCTAACC 1440

Query 1441  TTCGGGAGGACGGTTACCACGGTGTGATTCATGACTGGGG
            TGAAGTCGTAACAAGGTAGC 1500
            |||||||||||||||||||||||||||||||||||||||
            ||||||
Sbjct 1441  TTCGGGAGGACGGTTACCACGGTGTGATTCATGACTGGGG
            TGAAGTCGTAACAAGGTAGC 1500

Query 1501  CGTAGGGGAACCTGCGGCTGGATCACCTCCTT 1532
            ||||||||||||||||||||||||||||||||
Sbjct 1501  CGTAGGGGAACCTGCGGCTGGATCACCTCCTT 1532
```

Download GenBank Graphics

E value Score Percent identity Query start position Subject start position

Next Previous Descriptions

Pseudomonas fluorescens strain CFBP2392 chromosome, complete genome
CP133209.1 6656414 6

GenBank Graphics Next Match Previous Match
Alignment statistics for match #1

Score	Expect	Identities	Gaps	Strand
2824 bits(1529)	0.0	1531/1532(99%)	0/1532(0%)	Plus/Plus

Query 1   TGAAGAGTTTGATCATGGCTCAGATTGAACGCTGGCGGCA
GGCCTAACACATGCAAGTCG *60*
       |||||||||||||||||||||||||||||||||||||||||||||||||||||||
||||||

Sbjct 488   TGAAGAGTTTGATCATGGCTCAGATTGAACGCTGGCGGCA
GGCCTAACACATGCAAGTCG *547*

Query 61   AGCGGATGACAGGAGCTTGCTCCTGAATTCAGCGGCGGA
CGGGTGAGTAATGCCTAGGAA *120*
       |||||||||||||||||||||||||||||||||||||||||||||||||||||||
||||||

Sbjct 548   AGCGGATGACAGGAGCTTGCTCCTGAATTCAGCGGCGGA
CGGGTGAGTAATGCCTAGGAA *607*

Query 121   TCTGCCTGGTAGTGGGGGACAACGTTTCGAAAGGAACGC
TAATACCGCATACGTCCTACG *180*
       |||||||||||||||||||||||||||||||||||||||||||||||||||||||
||||||

Sbjct 608   TCTGCCTGGTAGTGGGGGACAACGTTTCGAAAGGAACGC
TAATACCGCATACGTCCTACG *667*

Query 181   GGAGAAAGCAGGGGACCTTCGGGCCTTGCGCTATCAGAT
GAGCCTAGGTCGGATTAGCTA *240*
       |||||||||||||||||||||||||||||||||||||||||||||||||||||||
||||||

Sbjct 668   GGAGAAAGCAGGGGACCTTCGGGCCTTGCGCTATCAGAT
GAGCCTAGGTCGGATTAGCTA *727*

```
Query 241   GTTGGTGAGGTAATGGCTCACCAAGGCGACGATCCGTAAC
            TGGTCTGAGAGGATGATCAG 300
            ||||||||||||||||||||||||||||||||||||||||
            ||||||
Sbjct 728   GTTGGTGAGGTAATGGCTCACCAAGGCGACGATCCGTAAC
            TGGTCTGAGAGGATGATCAG 787

Query 301   TCACACTGGAACTGAGACACGGTCCAGACTCCTACGGGA
            GGCAGCAGTGGGGAATATTGG 360
            ||||||||||||||||||||||||||||||||||||||||
            ||||||
Sbjct 788   TCACACTGGAACTGAGACACGGTCCAGACTCCTACGGGA
            GGCAGCAGTGGGGAATATTGG 847

Query 361   ACAATGGGCGAAAGCCTGATCCAGCCATGCCGCGTGTGTG
            AAGAAGGTCTTCGGATTGTA 420
            ||||||||||||||||||||||||||||||||||||||||
            ||||||
Sbjct 848   ACAATGGGCGAAAGCCTGATCCAGCCATGCCGCGTGTGTG
            AAGAAGGTCTTCGGATTGTA 907

Query 421   AAGCACTTTAAGTTGGGAGGAAGGGTTGTAGATTAATACT
            CTGCAATTTTGACGTTACCG 480
            ||||||||||||||||||||||||||||||||||||||||
            ||||||
Sbjct 908   AAGCACTTTAAGTTGGGAGGAAGGGTTGTAGATTAATACT
            CTGCAATTTTGACGTTACCG 967

Query 481   ACAGAATAAGCACCGGCTAACTCTGTGCCAGCAGCCGCG
            GTAATACAGAGGGTGCAAGCG 540
            ||||||||||||||||||||||||||||||||||||||||
            ||||||
Sbjct 968   ACAGAATAAGCACCGGCTAACTCTGTGCCAGCAGCCGCG
            GTAATACAGAGGGTGCAAGCG 1027

Query 541   TTAATCGGAATTACTGGGCGTAAAGCGCGCGTAGGTGGTT
            CGTTAAGTTGGATGTGAAAT 600
            ||||||||||||||||||||||||||||||||||||||||
            ||||||
Sbjct 1028  TTAATCGGAATTACTGGGCGTAAAGCGCGCGTAGGTGGTT
            CGTTAAGTTGGATGTGAAAT 1087
```

```
Query 601   CCCCGGGCTCAACCTGGGAACTGCATTCAAAACTGTCGAG
            CTAGAGTATGGTAGAGGGTG 660
            ||||||||||||||||||||||||||||||||||||||||
            ||||||
Sbjct 1088  CCCCGGGCTCAACCTGGGAACTGCATTCAAAACTGTCGA
            GCTAGAGTATGGTAGAGGGTG 1147

Query 661   GTGGAATTTCCTGTGTAGCGGTGAAATGCGTAGATATAGG
            AAGGAACACCAGTGGCGAAG 720
            ||||||||||||||||||||||||||||||||||||||||
            ||||||
Sbjct 1148  GTGGAATTTCCTGTGTAGCGGTGAAATGCGTAGATATAGG
            AAGGAACACCAGTGGCGAAG 1207

Query 721   GCGACCACCTGGACTGATACTGACACTGAGGTGCGAAAG
            CGTGGGGAGCAAACAGGATTA 780
            ||||||||||||||||||||||||||||||||||||||||
            ||||||
Sbjct 1208  GCGACCACCTGGACTGATACTGACACTGAGGTGCGAAA
            GCGTGGGGAGCAAACAGGATTA 1267

Query 781   GATACCCTGGTAGTCCACGCCGTAAACGATGTCAACTAGC
            CGTTGGGAGCCTTGAGCTCT 840
            ||||||||||||||||||||||||||||||||||||||||
            ||||||
Sbjct 1268  GATACCCTGGTAGTCCACGCCGTAAACGATGTCAACTAGC
            CGTTGGGAGCCTTGAGCTCT 1327

Query 841   TAGTGGCGCAGCTAACGCATTAAGTTGACCGCCTGGGGAG
            TACGGCCGCAAGGTTAAAAC 900
            ||||||||||||||||||||||||||||||||||||||||
            ||||||
Sbjct 1328  TAGTGGCGCAGCTAACGCATTAAGTTGACCGCCTGGGG
            AGTACGGCCGCAAGGTTAAAAC 1387

Query 901   TCAAATGAATTGACGGGGGCCCGCACAAGCGGTGGAGCA
            TGTGGTTTAATTCGAAGCAAC 960
            ||||||||||||||||||||||||||||||||||||||||
            ||||||
Sbjct 1388  TCAAATGAATTGACGGGGGCCCGCACAAGCGGTGGAGCA
            TGTGGTTTAATTCGAAGCAAC 1447
```

```
Query 961   GCGAAGAACCTTACCAGGCCTTGACATCCAATGAACTTTC
            CAGAGATGGATTGGTGCCTT 1020
            ||||||||||||||||||||||||||||||||||||||||
            ||||||
Sbjct 1448  GCGAAGAACCTTACCAGGCCTTGACATCCAATGAACTTTC
            CAGAGATGGATTGGTGCCTT 1507

Query 1021  CGGGAGCATTGAGACAGGTGCTGCATGGCTGTCGTCAGCT
            CGTGTCGTGAGATGTTGGGT 1080
            ||||| ||||||||||||||||||||||||||||||||||
            ||||||
Sbjct 1508  CGGGAACATTGAGACAGGTGCTGCATGGCTGTCGTCAGCT
            CGTGTCGTGAGATGTTGGGT 1567

Query 1081  TAAGTCCCGTAACGAGCGCAACCCTTGTCCTTAGTTACCA
            GCACGTTATGGTGGGCACTC 1140
            ||||||||||||||||||||||||||||||||||||||||
            ||||||
Sbjct 1568  TAAGTCCCGTAACGAGCGCAACCCTTGTCCTTAGTTACCA
            GCACGTTATGGTGGGCACTC 1627

Query 1141  TAAGGAGACTGCCGGTGACAAACCGGAGGAAGGTGGGGA
            TGACGTCAAGTCATCATGGCC 1200
            ||||||||||||||||||||||||||||||||||||||||
            ||||||
Sbjct 1628  TAAGGAGACTGCCGGTGACAAACCGGAGGAAGGTGGGGA
            TGACGTCAAGTCATCATGGCC 1687

Query 1201  CTTACGGCCTGGGCTACACACGTGCTACAATGGTCGGTAC
            AAAGGGTTGCCAAGCCGCGA 1260
            ||||||||||||||||||||||||||||||||||||||||
            ||||||
Sbjct 1688  CTTACGGCCTGGGCTACACACGTGCTACAATGGTCGGTAC
            AAAGGGTTGCCAAGCCGCGA 1747

Query 1261  GGTGGAGCTAATCCCATAAAACCGATCGTAGTCCGGATCG
            CAGTCTGCAACTCGACTGCG 1320
            ||||||||||||||||||||||||||||||||||||||||
            ||||||
Sbjct 1748  GGTGGAGCTAATCCCATAAAACCGATCGTAGTCCGGATCG
            CAGTCTGCAACTCGACTGCG 1807
```

Query 1321 TGAAGTCGGAATCGCTAGTAATCGTGAATCAGAATGTCAC
GGTGAATACGTTCCCGGGCC 1380
            ||||||||||||||||||||||||||||||||||||||||
            ||||||
Sbjct 1808 TGAAGTCGGAATCGCTAGTAATCGTGAATCAGAATGTCAC
GGTGAATACGTTCCCGGGCC 1867

Query 1381 TTGTACACACCGCCCGTCACACCATGGGAGTGGGTTGCAC
CAGAAGTAGCTAGTCTAACC 1440
            ||||||||||||||||||||||||||||||||||||||||
            ||||||
Sbjct 1868 TTGTACACACCGCCCGTCACACCATGGGAGTGGGTTGCAC
CAGAAGTAGCTAGTCTAACC 1927

Query 1441 TTCGGGAGGACGGTTACCACGGTGTGATTCATGACTGGGG
TGAAGTCGTAACAAGGTAGC 1500
            ||||||||||||||||||||||||||||||||||||||||
            ||||||
Sbjct 1928 TTCGGGAGGACGGTTACCACGGTGTGATTCATGACTGGGG
TGAAGTCGTAACAAGGTAGC 1987

Query 1501 CGTAGGGGAACCTGCGGCTGGATCACCTCCTT 1532
            ||||||||||||||||||||||||||||||||
Sbjct 1988 CGTAGGGGAACCTGCGGCTGGATCACCTCCTT 2019

GenBank Graphics Next Match Previous Match First Match
Alignment statistics for match #2

Score	Expect	Identities	Gaps	Strand
2824 bits(1529)	0.0	1531/1532(99%)	0/1532(0%)	Plus/Plus

Query 1   TGAAGAGTTTGATCATGGCTCAGATTGAACGCTGGCGGC
AGGCCTAACACATGCAAGTCG 60
              |||||||||||||||||||||||||||||||||||||||
              |||||||||
Sbjct 707876 TGAAGAGTTTGATCATGGCTCAGATTGAACGCTGGCGGC
AGGCCTAACACATGCAAGTCG 707935

Query 61   AGCGGATGACAGGAGCTTGCTCCTGAATTCAGCGGCGG
ACGGGTGAGTAATGCCTAGGAA 120
              |||||||||||||||||||||||||||||||||||||||
              |||||||||
Sbjct 707936 AGCGGATGACAGGAGCTTGCTCCTGAATTCAGCGGCGG
ACGGGTGAGTAATGCCTAGGAA 707995

Query 121   TCTGCCTGGTAGTGGGGGACAACGTTTCGAAAGGAACG
            CTAATACCGCATACGTCCTACG 180
            |||||||||||||||||||||||||||||||||||||||
            |||||||||
Sbjct 707996 TCTGCCTGGTAGTGGGGGACAACGTTTCGAAAGGAACG
            CTAATACCGCATACGTCCTACG 708055

Query 181   GGAGAAAGCAGGGGACCTTCGGGCCTTGCGCTATCAGA
            TGAGCCTAGGTCGGATTAGCTA 240
            |||||||||||||||||||||||||||||||||||||||
            |||||||||
Sbjct 708056 GGAGAAAGCAGGGGACCTTCGGGCCTTGCGCTATCAGA
            TGAGCCTAGGTCGGATTAGCTA 708115

Query 241   GTTGGTGAGGTAATGGCTCACCAAGGCGACGATCCGTAA
            CTGGTCTGAGAGGATGATCAG 300
            |||||||||||||||||||||||||||||||||||||||
            |||||||||
Sbjct 708116 GTTGGTGAGGTAATGGCTCACCAAGGCGACGATCCGTAA
            CTGGTCTGAGAGGATGATCAG 708175

Query 301   TCACACTGGAACTGAGACACGGTCCAGACTCCTACGGG
            AGGCAGCAGTGGGGAATATTGG 360
            |||||||||||||||||||||||||||||||||||||||
            |||||||||
Sbjct 708176 TCACACTGGAACTGAGACACGGTCCAGACTCCTACGGG
            AGGCAGCAGTGGGGAATATTGG 708235

Query 361   ACAATGGGCGAAAGCCTGATCCAGCCATGCCGCGTGTG
            TGAAGAAGGTCTTCGGATTGTA 420
            |||||||||||||||||||||||||||||||||||||||
            |||||||||
Sbjct 708236 ACAATGGGCGAAAGCCTGATCCAGCCATGCCGCGTGTG
            TGAAGAAGGTCTTCGGATTGTA 708295

Query 421   AAGCACTTTAAGTTGGGAGGAAGGGTTGTAGATTAATAC
            TCTGCAATTTTGACGTTACCG 480
            |||||||||||||||||||||||||||||||||||||||
            |||||||||
Sbjct 708296 AAGCACTTTAAGTTGGGAGGAAGGGTTGTAGATTAATAC
            TCTGCAATTTTGACGTTACCG 708355

Query 481   ACAGAATAAGCACCGGCTAACTCTGTGCCAGCAGCCGC
             GGTAATACAGAGGGTGCAAGCG 540
             ||||||||||||||||||||||||||||||||||||||||
             |||||||||
Sbjct 708356 ACAGAATAAGCACCGGCTAACTCTGTGCCAGCAGCCGC
             GGTAATACAGAGGGTGCAAGCG 708415

Query 541   TTAATCGGAATTACTGGGCGTAAAGCGCGCGTAGGTGGT
             TCGTTAAGTTGGATGTGAAAT 600
             ||||||||||||||||||||||||||||||||||||||||
             |||||||||
Sbjct 708416 TTAATCGGAATTACTGGGCGTAAAGCGCGCGTAGGTGGT
             TCGTTAAGTTGGATGTGAAAT 708475

Query 601   CCCCGGGCTCAACCTGGGAACTGCATTCAAAACTGTCGA
             GCTAGAGTATGGTAGAGGGTG 660
             ||||||||||||||||||||||||||||||||||||||||
             |||||||||
Sbjct 708476 CCCCGGGCTCAACCTGGGAACTGCATTCAAAACTGTCGA
             GCTAGAGTATGGTAGAGGGTG 708535

Query 661   GTGGAATTTCCTGTGTAGCGGTGAAATGCGTAGATATAG
             GAAGGAACACCAGTGGCGAAG 720
             ||||||||||||||||||||||||||||||||||||||||
             |||||||||
Sbjct 708536 GTGGAATTTCCTGTGTAGCGGTGAAATGCGTAGATATAG
             GAAGGAACACCAGTGGCGAAG 708595

Query 721   GCGACCACCTGGACTGATACTGACACTGAGGTGCGAAA
             GCGTGGGGAGCAAACAGGATTA 780
             ||||||||||||||||||||||||||||||||||||||||
             |||||||||
Sbjct 708596 GCGACCACCTGGACTGATACTGACACTGAGGTGCGAAA
             GCGTGGGGAGCAAACAGGATTA 708655

Query 781   GATACCCTGGTAGTCCACGCCGTAAACGATGTCAACTAG
             CCGTTGGGAGCCTTGAGCTCT 840
             ||||||||||||||||||||||||||||||||||||||||
             |||||||||
Sbjct 708656 GATACCCTGGTAGTCCACGCCGTAAACGATGTCAACTAG
             CCGTTGGGAGCCTTGAGCTCT 708715

Query 841   TAGTGGCGCAGCTAACGCATTAAGTTGACCGCCTGGGGA
             GTACGGCCGCAAGGTTAAAAC 900
             |||||||||||||||||||||||||||||||||||||||
             |||||||||
Sbjct 708716 TAGTGGCGCAGCTAACGCATTAAGTTGACCGCCTGGGGA
             GTACGGCCGCAAGGTTAAAAC 708775

Query 901   TCAAATGAATTGACGGGGGCCCGCACAAGCGGTGGAGC
             ATGTGGTTTAATTCGAAGCAAC 960
             |||||||||||||||||||||||||||||||||||||||
             |||||||||
Sbjct 708776 TCAAATGAATTGACGGGGGCCCGCACAAGCGGTGGAGC
             ATGTGGTTTAATTCGAAGCAAC 708835

Query 961   GCGAAGAACCTTACCAGGCCTTGACATCCAATGAACTTT
             CCAGAGATGGATTGGTGCCTT 1020
             |||||||||||||||||||||||||||||||||||||||
             |||||||||
Sbjct 708836 GCGAAGAACCTTACCAGGCCTTGACATCCAATGAACTTT
             CCAGAGATGGATTGGTGCCTT 708895

Query 1021   CGGGAGCATTGAGACAGGTGCTGCATGGCTGTCGTCAG
             CTCGTGTCGTGAGATGTTGGGT 1080
             |||||||||||||||||||||||||||||||||||||||
             |||||||||
Sbjct 708896 CGGGAGCATTGAGACAGGTGCTGCATGGCTGTCGTCAG
             CTCGTGTCGTGAGATGTTGGGT 708955

Query 1081   TAAGTCCCGTAACGAGCGCAACCCTTGTCCTTAGTTACC
             AGCACGTTATGGTGGGCACTC 1140
             |||||||||||||||||||||||||||||||||||||| ||||||
             |||||||
Sbjct 708956 TAAGTCCCGTAACGAGCGCAACCCTTGTCCTTAGTTACC
             AGCACGTCATGGTGGGCACTC 709015

Query 1141   TAAGGAGACTGCCGGTGACAAACCGGAGGAAGGTGGG
             GATGACGTCAAGTCATCATGGCC 1200
             |||||||||||||||||||||||||||||||||||||||
             |||||||||
Sbjct 709016 TAAGGAGACTGCCGGTGACAAACCGGAGGAAGGTGGG
             GATGACGTCAAGTCATCATGGCC 709075

Query 1201 CTTACGGCCTGGGCTACACACGTGCTACAATGGTCGGTA
CAAAGGGTTGCCAAGCCGCGA 1260
||||||||||||||||||||||||||||||||||||||||
||||||||
Sbjct 709076 CTTACGGCCTGGGCTACACACGTGCTACAATGGTCGGTA
CAAAGGGTTGCCAAGCCGCGA 709135

Query 1261 GGTGGAGCTAATCCCATAAAACCGATCGTAGTCCGGATC
GCAGTCTGCAACTCGACTGCG 1320
||||||||||||||||||||||||||||||||||||||||
||||||||
Sbjct 709136 GGTGGAGCTAATCCCATAAAACCGATCGTAGTCCGGATC
GCAGTCTGCAACTCGACTGCG 709195

Query 1321 TGAAGTCGGAATCGCTAGTAATCGTGAATCAGAATGTCA
CGGTGAATACGTTCCCGGGCC 1380
||||||||||||||||||||||||||||||||||||||||
||||||||
Sbjct 709196 TGAAGTCGGAATCGCTAGTAATCGTGAATCAGAATGTCA
CGGTGAATACGTTCCCGGGCC 709255

Query 1381 TTGTACACACCGCCCGTCACACCATGGGAGTGGGTTGCA
CCAGAAGTAGCTAGTCTAACC 1440
||||||||||||||||||||||||||||||||||||||||
||||||||
Sbjct 709256 TTGTACACACCGCCCGTCACACCATGGGAGTGGGTTGCA
CCAGAAGTAGCTAGTCTAACC 709315

Query 1441 TTCGGGAGGACGGTTACCACGGTGTGATTCATGACTGGG
GTGAAGTCGTAACAAGGTAGC 1500
||||||||||||||||||||||||||||||||||||||||
||||||||
Sbjct 709316 TTCGGGAGGACGGTTACCACGGTGTGATTCATGACTGGG
GTGAAGTCGTAACAAGGTAGC 709375

Query 1501 CGTAGGGGAACCTGCGGCTGGATCACCTCCTT 1532
||||||||||||||||||||||||||||||||
Sbjct 709376 CGTAGGGGAACCTGCGGCTGGATCACCTCCTT 709407

GenBank Graphics Next Match Previous Match First Match
Alignment statistics for match #3

Score	Expect	Identities	Gaps	Strand
2824 bits(1529)	0.0	1531/1532(99%)	0/1532(0%)	Plus/Minus

```
Query 1         TGAAGAGTTTGATCATGGCTCAGATTGAACGCTGGCGG
                CAGGCCTAACACATGCAAGTCG 60
                ||||||||||||||||||||||||||||||||||||||
                |||||||||
Sbjct 4500550   TGAAGAGTTTGATCATGGCTCAGATTGAACGCTGGCGG
                CAGGCCTAACACATGCAAGTCG 4500491

Query 61        AGCGGATGACAGGAGCTTGCTCCTGAATTCAGCGGCG
                GACGGGTGAGTAATGCCTAGGAA 120
                ||||||||||||||||||||||||||||||||||||||
                |||||||||
Sbjct 4500490   AGCGGATGACAGGAGCTTGCTCCTGAATTCAGCGGCG
                GACGGGTGAGTAATGCCTAGGAA 4500431

Query 121       TCTGCCTGGTAGTGGGGGACAACGTTTCGAAAGGAAC
                GCTAATACCGCATACGTCCTACG 180
                ||||||||||||||||||||||||||||||||||||||
                |||||||||
Sbjct 4500430   TCTGCCTGGTAGTGGGGGACAACGTTTCGAAAGGAAC
                GCTAATACCGCATACGTCCTACG 4500371

Query 181       GGAGAAAGCAGGGGACCTTCGGGCCTTGCGCTATCAG
                ATGAGCCTAGGTCGGATTAGCTA 240
                ||||||||||||||||||||||||||||||||||||||
                |||||||||
Sbjct 4500370   GGAGAAAGCAGGGGACCTTCGGGCCTTGCGCTATCAG
                ATGAGCCTAGGTCGGATTAGCTA 4500311

Query 241       GTTGGTGAGGTAATGGCTCACCAAGGCGACGATCCGTA
                ACTGGTCTGAGAGGATGATCAG 300
                ||||||||||||||||||||||||||||||||||||||
                |||||||||
Sbjct 4500310   GTTGGTGAGGTAATGGCTCACCAAGGCGACGATCCGTA
                ACTGGTCTGAGAGGATGATCAG 4500251

Query 301       TCACACTGGAACTGAGACACGGTCCAGACTCCTACGG
                GAGGCAGCAGTGGGGAATATTGG 360
                ||||||||||||||||||||||||||||||||||||||
                |||||||||
Sbjct 4500250   TCACACTGGAACTGAGACACGGTCCAGACTCCTACGG
                GAGGCAGCAGTGGGGAATATTGG 4500191
```

Query 361   ACAATGGGCGAAAGCCTGATCCAGCCATGCCGCGTGT
               GTGAAGAAGGTCTTCGGATTGTA 420
               |||||||||||||||||||||||||||||||||||||
               ||||||||||
Sbjct 4500190 ACAATGGGCGAAAGCCTGATCCAGCCATGCCGCGTGT
               GTGAAGAAGGTCTTCGGATTGTA 4500131

Query 421   AAGCACTTTAAGTTGGGAGGAAGGGTTGTAGATTAATA
               CTCTGCAATTTTGACGTTACCG 480
               |||||||||||||||||||||||||||||||||||||
               ||||||||||
Sbjct 4500130 AAGCACTTTAAGTTGGGAGGAAGGGTTGTAGATTAATA
               CTCTGCAATTTTGACGTTACCG 4500071

Query 481   ACAGAATAAGCACCGGCTAACTCTGTGCCAGCAGCCG
               CGGTAATACAGAGGGTGCAAGCG 540
               |||||||||||||||||||||||||||||||||||||
               ||||||||||
Sbjct 4500070 ACAGAATAAGCACCGGCTAACTCTGTGCCAGCAGCCG
               CGGTAATACAGAGGGTGCAAGCG 4500011

Query 541   TTAATCGGAATTACTGGGCGTAAAGCGCGCGTAGGTGG
               TTCGTTAAGTTGGATGTGAAAT 600
               |||||||||||||||||||||||||||||||||||||
               ||||||||||
Sbjct 4500010 TTAATCGGAATTACTGGGCGTAAAGCGCGCGTAGGTGG
               TTCGTTAAGTTGGATGTGAAAT 4499951

Query 601   CCCCGGGCTCAACCTGGGAACTGCATTCAAAACTGTC
               GAGCTAGAGTATGGTAGAGGGTG 660
               |||||||||||||||||||||||||||||||||||||
               ||||||||||
Sbjct 4499950 CCCCGGGCTCAACCTGGGAACTGCATTCAAAACTGTC
               GAGCTAGAGTATGGTAGAGGGTG 4499891

Query 661   GTGGAATTTCCTGTGTAGCGGTGAAATGCGTAGATATA
               GGAAGGAACACCAGTGGCGAAG 720
               |||||||||||||||||||||||||||||||||||||
               ||||||||||
Sbjct 4499890 GTGGAATTTCCTGTGTAGCGGTGAAATGCGTAGATATA
               GGAAGGAACACCAGTGGCGAAG 4499831

```
Query 721       GCGACCACCTGGACTGATACTGACACTGAGGTGCGAA
                AGCGTGGGGAGCAAACAGGATTA 780
                |||||||||||||||||||||||||||||||||||||
                ||||||||||
Sbjct 4499830   GCGACCACCTGGACTGATACTGACACTGAGGTGCGAA
                AGCGTGGGGAGCAAACAGGATTA 4499771

Query 781       GATACCCTGGTAGTCCACGCCGTAAACGATGTCAACTA
                GCCGTTGGGAGCCTTGAGCTCT 840
                |||||||||||||||||||||||||||||||||||||||
                ||||||||||
Sbjct 4499770   GATACCCTGGTAGTCCACGCCGTAAACGATGTCAACTA
                GCCGTTGGGAGCCTTGAGCTCT 4499711

Query 841       TAGTGGCGCAGCTAACGCATTAAGTTGACCGCCTGGG
                GAGTACGGCCGCAAGGTTAAAAC 900
                |||||||||||||||||||||||||||||||||||||||
                ||||||||||
Sbjct 4499710   TAGTGGCGCAGCTAACGCATTAAGTTGACCGCCTGGG
                GAGTACGGCCGCAAGGTTAAAAC 4499651

Query 901       TCAAATGAATTGACGGGGGCCCGCACAAGCGGTGGAG
                CATGTGGTTTAATTCGAAGCAAC 960
                |||||||||||||||||||||||||||||||||||||||
                ||||||||||
Sbjct 4499650   TCAAATGAATTGACGGGGGCCCGCACAAGCGGTGGAG
                CATGTGGTTTAATTCGAAGCAAC 4499591

Query 961       GCGAAGAACCTTACCAGGCCTTGACATCCAATGAACT
                TTCCAGAGATGGATTGGTGCCTT 1020
                |||||||||||||||||||||||||||||||||||||||
                ||||||||||
Sbjct 4499590   GCGAAGAACCTTACCAGGCCTTGACATCCAATGAACTT
                TCCAGAGATGGATTGGTGCCTT 4499531

Query 1021      CGGGAGCATTGAGACAGGTGCTGCATGGCTGTCGTCA
                GCTCGTGTCGTGAGATGTTGGGT 1080
                ||||| ||||||||||||||||||||||||||||||||
                ||||||||||
Sbjct 4499530   CGGGAACATTGAGACAGGTGCTGCATGGCTGTCGTCA
                GCTCGTGTCGTGAGATGTTGGGT 4499471
```

Query 1081   TAAGTCCCGTAACGAGCGCAACCCTTGTCCTTAGTTAC
CAGCACGTTATGGTGGGCACTC 1140
              | | | | | | | | | | | | | | | | | | | | | | | | | | | | | | | | | | | | | | | | | |
              | | | | | | | | |
Sbjct 4499470   TAAGTCCCGTAACGAGCGCAACCCTTGTCCTTAGTTAC
CAGCACGTTATGGTGGGCACTC 4499411

Query 1141   TAAGGAGACTGCCGGTGACAAACCGGAGGAAGGTGG
GGATGACGTCAAGTCATCATGGCC 1200
              | | | | | | | | | | | | | | | | | | | | | | | | | | | | | | | | | | | | | | | | | |
              | | | | | | | | |
Sbjct 4499410   TAAGGAGACTGCCGGTGACAAACCGGAGGAAGGTGG
GGATGACGTCAAGTCATCATGGCC 4499351

Query 1201   CTTACGGCCTGGGCTACACACGTGCTACAATGGTCGGT
ACAAAGGGTTGCCAAGCCGCGA 1260
              | | | | | | | | | | | | | | | | | | | | | | | | | | | | | | | | | | | | | | | | | |
              | | | | | | | | |
Sbjct 4499350   CTTACGGCCTGGGCTACACACGTGCTACAATGGTCGGT
ACAAAGGGTTGCCAAGCCGCGA 4499291

Query 1261   GGTGGAGCTAATCCCATAAAACCGATCGTAGTCCGGAT
CGCAGTCTGCAACTCGACTGCG 1320
              | | | | | | | | | | | | | | | | | | | | | | | | | | | | | | | | | | | | | | | | | |
              | | | | | | | | |
Sbjct 4499290   GGTGGAGCTAATCCCATAAAACCGATCGTAGTCCGGAT
CGCAGTCTGCAACTCGACTGCG 4499231

Query 1321   TGAAGTCGGAATCGCTAGTAATCGTGAATCAGAATGTC
ACGGTGAATACGTTCCCGGGCC 1380
              | | | | | | | | | | | | | | | | | | | | | | | | | | | | | | | | | | | | | | | | | |
              | | | | | | | | |
Sbjct 4499230   TGAAGTCGGAATCGCTAGTAATCGTGAATCAGAATGTC
ACGGTGAATACGTTCCCGGGCC 4499171

Query 1381   TTGTACACACCGCCCGTCACACCATGGGAGTGGGTTGC
ACCAGAAGTAGCTAGTCTAACC 1440
              | | | | | | | | | | | | | | | | | | | | | | | | | | | | | | | | | | | | | | | | | |
              | | | | | | | | |
Sbjct 4499170   TTGTACACACCGCCCGTCACACCATGGGAGTGGGTTGC
ACCAGAAGTAGCTAGTCTAACC 4499111

Query 1441   TTCGGGAGGACGGTTACCACGGTGTGATTCATGACTGG
              GGTGAAGTCGTAACAAGGTAGC 1500
              |||||||||||||||||||||||||||||||||||||||||||||||||
              |||||||||
Sbjct 4499110 TTCGGGAGGACGGTTACCACGGTGTGATTCATGACTGG
              GGTGAAGTCGTAACAAGGTAGC 4499051

Query 1501   CGTAGGGGAACCTGCGGCTGGATCACCTCCTT 1532
              ||||||||||||||||||||||||||||||||
Sbjct 4499050 CGTAGGGGAACCTGCGGCTGGATCACCTCCTT 4499019

GenBank Graphics Next Match Previous Match First Match
Alignment statistics for match #4

Score	Expect	Identities	Gaps	Strand
2824 bits(1529)	0.0	1531/1532(99%)	0/1532(0%)	Plus/Minus

Query 1   TGAAGAGTTTGATCATGGCTCAGATTGAACGCTGGCGG
             CAGGCCTAACACATGCAAGTCG 60
             |||||||||||||||||||||||||||||||||||||||||||||||||
             |||||||||
Sbjct 4892727 TGAAGAGTTTGATCATGGCTCAGATTGAACGCTGGCGG
             CAGGCCTAACACATGCAAGTCG 4892668

Query 61   AGCGGATGACAGGAGCTTGCTCCTGAATTCAGCGGCG
             GACGGGTGAGTAATGCCTAGGAA 120
             |||||||||||||||||||||||||||||||||||||||||||||||||
             |||||||||
Sbjct 4892667 AGCGGATGACAGGAGCTTGCTCCTGAATTCAGCGGCG
             GACGGGTGAGTAATGCCTAGGAA 4892608

Query 121   TCTGCCTGGTAGTGGGGGACAACGTTTCGAAAGGAAC
             GCTAATACCGCATACGTCCTACG 180
             |||||||||||||||||||||||||||||||||||||||||||||||||
             |||||||||
Sbjct 4892607 TCTGCCTGGTAGTGGGGGACAACGTTTCGAAAGGAAC
             GCTAATACCGCATACGTCCTACG 4892548

Query 181   GGAGAAAGCAGGGGACCTTCGGGCCTTGCGCTATCAG
             ATGAGCCTAGGTCGGATTAGCTA 240
             |||||||||||||||||||||||||||||||||||||||||||||||||
             |||||||||
Sbjct 4892547 GGAGAAAGCAGGGGACCTTCGGGCCTTGCGCTATCAG
             ATGAGCCTAGGTCGGATTAGCTA 4892488

```
Query 241    GTTGGTGAGGTAATGGCTCACCAAGGCGACGATCCGTA
             ACTGGTCTGAGAGGATGATCAG 300
             ||||||||||||||||||||||||||||||||||||||||||||||||
             |||||||||
Sbjct 4892487  GTTGGTGAGGTAATGGCTCACCAAGGCGACGATCCGT
             AACTGGTCTGAGAGGATGATCAG 4892428

Query 301    TCACACTGGAACTGAGACACGGTCCAGACTCCTACGG
             GAGGCAGCAGTGGGGAATATTGG 360
             ||||||||||||||||||||||||||||||||||||||||||||||||
             |||||||||
Sbjct 4892427  TCACACTGGAACTGAGACACGGTCCAGACTCCTACGG
             GAGGCAGCAGTGGGGAATATTGG 4892368

Query 361    ACAATGGGCGAAAGCCTGATCCAGCCATGCCGCGTGT
             GTGAAGAAGGTCTTCGGATTGTA 420
             ||||||||||||||||||||||||||||||||||||||||||||||||
             |||||||||
Sbjct 4892367  ACAATGGGCGAAAGCCTGATCCAGCCATGCCGCGTGT
             GTGAAGAAGGTCTTCGGATTGTA 4892308

Query 421    AAGCACTTTAAGTTGGGAGGAAGGGTTGTAGATTAATA
             CTCTGCAATTTTGACGTTACCG 480
             ||||||||||||||||||||||||||||||||||||||||||||||||
             |||||||||
Sbjct 4892307  AAGCACTTTAAGTTGGGAGGAAGGGTTGTAGATTAATA
             CTCTGCAATTTTGACGTTACCG 4892248

Query 481    ACAGAATAAGCACCGGCTAACTCTGTGCCAGCAGCCG
             CGGTAATACAGAGGGTGCAAGCG 540
             ||||||||||||||||||||||||||||||||||||||||||||||||
             |||||||||
Sbjct 4892247  ACAGAATAAGCACCGGCTAACTCTGTGCCAGCAGCCG
             CGGTAATACAGAGGGTGCAAGCG 4892188

Query 541    TTAATCGGAATTACTGGGCGTAAAGCGCGCGTAGGTGG
             TTCGTTAAGTTGGATGTGAAAT 600
             ||||||||||||||||||||||||||||||||||||||||||||||||
             |||||||||
Sbjct 4892187  TTAATCGGAATTACTGGGCGTAAAGCGCGCGTAGGTGG
             TTCGTTAAGTTGGATGTGAAAT 4892128
```

Query 601   CCCCGGGCTCAACCTGGGAACTGCATTCAAAACTGTC
            GAGCTAGAGTATGGTAGAGGGTG 660
            |||||||||||||||||||||||||||||||||||||
            |||||||||
Sbjct 4892127 CCCCGGGCTCAACCTGGGAACTGCATTCAAAACTGTC
            GAGCTAGAGTATGGTAGAGGGTG 4892068

Query 661   GTGGAATTTCCTGTGTAGCGGTGAAATGCGTAGATATA
            GGAAGGAACACCAGTGGCGAAG 720
            |||||||||||||||||||||||||||||||||||||
            |||||||||
Sbjct 4892067 GTGGAATTTCCTGTGTAGCGGTGAAATGCGTAGATATA
            GGAAGGAACACCAGTGGCGAAG 4892008

Query 721   GCGACCACCTGGACTGATACTGACACTGAGGTGCGAA
            AGCGTGGGGAGCAAACAGGATTA 780
            |||||||||||||||||||||||||||||||||||||
            |||||||||
Sbjct 4892007 GCGACCACCTGGACTGATACTGACACTGAGGTGCGAA
            AGCGTGGGGAGCAAACAGGATTA 4891948

Query 781   GATACCCTGGTAGTCCACGCCGTAAACGATGTCAACTA
            GCCGTTGGGAGCCTTGAGCTCT 840
            |||||||||||||||||||||||||||||||||||||
            |||||||||
Sbjct 4891947 GATACCCTGGTAGTCCACGCCGTAAACGATGTCAACTA
            GCCGTTGGGAGCCTTGAGCTCT 4891888

Query 841   TAGTGGCGCAGCTAACGCATTAAGTTGACCGCCTGGG
            GAGTACGGCCGCAAGGTTAAAAC 900
            |||||||||||||||||||||||||||||||||||||
            |||||||||
Sbjct 4891887 TAGTGGCGCAGCTAACGCATTAAGTTGACCGCCTGGG
            GAGTACGGCCGCAAGGTTAAAAC 4891828

Query 901   TCAAATGAATTGACGGGGGCCCGCACAAGCGGTGGAG
            CATGTGGTTTAATTCGAAGCAAC 960
            |||||||||||||||||||||||||||||||||||||
            |||||||||
Sbjct 4891827 TCAAATGAATTGACGGGGGCCCGCACAAGCGGTGGAG
            CATGTGGTTTAATTCGAAGCAAC 4891768

```
Query 961      GCGAAGAACCTTACCAGGCCTTGACATCCAATGAACT
               TTCCAGAGATGGATTGGTGCCTT 1020
               |||||||||||||||||||||||||||||||||||||
               ||||||||||
Sbjct 4891767  GCGAAGAACCTTACCAGGCCTTGACATCCAATGAACT
               TTCCAGAGATGGATTGGTGCCTT 4891708

Query 1021     CGGGAGCATTGAGACAGGTGCTGCATGGCTGTCGTCA
               GCTCGTGTCGTGAGATGTTGGGT 1080
               ||||| |||||||||||||||||||||||||||||||
               ||||||||||
Sbjct 4891707  CGGGAACATTGAGACAGGTGCTGCATGGCTGTCGTCA
               GCTCGTGTCGTGAGATGTTGGGT 4891648

Query 1081     TAAGTCCCGTAACGAGCGCAACCCTTGTCCTTAGTTAC
               CAGCACGTTATGGTGGGCACTC 1140
               |||||||||||||||||||||||||||||||||||||
               ||||||||||
Sbjct 4891647  TAAGTCCCGTAACGAGCGCAACCCTTGTCCTTAGTTAC
               CAGCACGTTATGGTGGGCACTC 4891588

Query 1141     TAAGGAGACTGCCGGTGACAAACCGGAGGAAGGTGG
               GGATGACGTCAAGTCATCATGGCC 1200
               |||||||||||||||||||||||||||||||||||||
               ||||||||||
Sbjct 4891587  TAAGGAGACTGCCGGTGACAAACCGGAGGAAGGTGG
               GGATGACGTCAAGTCATCATGGCC 4891528

Query 1201     CTTACGGCCTGGGCTACACACGTGCTACAATGGTCGGT
               ACAAAGGGTTGCCAAGCCGCGA 1260
               |||||||||||||||||||||||||||||||||||||
               ||||||||||
Sbjct 4891527  CTTACGGCCTGGGCTACACACGTGCTACAATGGTCGGT
               ACAAAGGGTTGCCAAGCCGCGA 4891468

Query 1261     GGTGGAGCTAATCCCATAAAACCGATCGTAGTCCGGAT
               CGCAGTCTGCAACTCGACTGCG 1320
               |||||||||||||||||||||||||||||||||||||
               ||||||||||
Sbjct 4891467  GGTGGAGCTAATCCCATAAAACCGATCGTAGTCCGGAT
               CGCAGTCTGCAACTCGACTGCG 4891408
```

Query 1321   TGAAGTCGGAATCGCTAGTAATCGTGAATCAGAATGTC
ACGGTGAATACGTTCCCGGGCC 1380
              |||||||||||||||||||||||||||||||||||||||||||||||||
              |||||||||
Sbjct 4891407 TGAAGTCGGAATCGCTAGTAATCGTGAATCAGAATGTC
ACGGTGAATACGTTCCCGGGCC 4891348

Query 1381   TTGTACACACCGCCCGTCACACCATGGGAGTGGGTTGC
ACCAGAAGTAGCTAGTCTAACC 1440
              |||||||||||||||||||||||||||||||||||||||||||||||||
              |||||||||
Sbjct 4891347 TTGTACACACCGCCCGTCACACCATGGGAGTGGGTTG
CACCAGAAGTAGCTAGTCTAACC 4891288

Query 1441   TTCGGGAGGACGGTTACCACGGTGTGATTCATGACTG
GGGTGAAGTCGTAACAAGGTAGC 1500
              |||||||||||||||||||||||||||||||||||||||||||||||||
              |||||||||
Sbjct 4891287 TTCGGGAGGACGGTTACCACGGTGTGATTCATGACTG
GGGTGAAGTCGTAACAAGGTAGC 4891228

Query 1501   CGTAGGGGAACCTGCGGCTGGATCACCTCCTT 1532
              |||||||||||||||||||||||||||||||||
Sbjct 4891227 CGTAGGGGAACCTGCGGCTGGATCACCTCCTT 4891196

GenBank Graphics Next Match Previous Match First Match
Alignment statistics for match #5

Score	Expect	Identities	Gaps	Strand
2824 bits(1529)	0.0	1531/1532(99%)	0/1532(0%)	Plus/Minus

Query 1   TGAAGAGTTTGATCATGGCTCAGATTGAACGCTGGCGG
CAGGCCTAACACATGCAAGTCG 60
             |||||||||||||||||||||||||||||||||||||||||||||||||||
             |||||||||
Sbjct 5845502 TGAAGAGTTTGATCATGGCTCAGATTGAACGCTGGCGG
CAGGCCTAACACATGCAAGTCG 5845443

Query 61   AGCGGATGACAGGAGCTTGCTCCTGAATTCAGCGGCG
GACGGGTGAGTAATGCCTAGGAA 120
             |||||||||||||||||||||||||||||||||||||||||||||||||||
             |||||||||
Sbjct 5845442 AGCGGATGACAGGAGCTTGCTCCTGAATTCAGCGGCG
GACGGGTGAGTAATGCCTAGGAA 5845383

Query 121   TCTGCCTGGTAGTGGGGGACAACGTTTCGAAAGGAAC
GCTAATACCGCATACGTCCTACG *180*
|||||||||||||||||||||||||||||||||||||||||||||||||||
|||||||||||
Sbjct 5845382   TCTGCCTGGTAGTGGGGGACAACGTTTCGAAAGGAAC
GCTAATACCGCATACGTCCTACG *5845323*

Query 181   GGAGAAAGCAGGGGACCTTCGGGCCTTGCGCTATCAG
ATGAGCCTAGGTCGGATTAGCTA *240*
|||||||||||||||||||||||||||||||||||||||||||||||||||
|||||||||||
Sbjct 5845322   GGAGAAAGCAGGGGACCTTCGGGCCTTGCGCTATCAG
ATGAGCCTAGGTCGGATTAGCTA *5845263*

Query 241   GTTGGTGAGGTAATGGCTCACCAAGGCGACGATCCGTA
ACTGGTCTGAGAGGATGATCAG *300*
|||||||||||||||||||||||||||||||||||||||||||||||||||
|||||||||||
Sbjct 5845262   GTTGGTGAGGTAATGGCTCACCAAGGCGACGATCCGTA
ACTGGTCTGAGAGGATGATCAG *5845203*

Query 301   TCACACTGGAACTGAGACACGGTCCAGACTCCTACGG
GAGGCAGCAGTGGGGAATATTGG *360*
|||||||||||||||||||||||||||||||||||||||||||||||||||
|||||||||||
Sbjct 5845202   TCACACTGGAACTGAGACACGGTCCAGACTCCTACGG
GAGGCAGCAGTGGGGAATATTGG *5845143*

Query 361   ACAATGGGCGAAAGCCTGATCCAGCCATGCCGCGTGT
GTGAAGAAGGTCTTCGGATTGTA *420*
|||||||||||||||||||||||||||||||||||||||||||||||||||
|||||||||||
Sbjct 5845142   ACAATGGGCGAAAGCCTGATCCAGCCATGCCGCGTGT
GTGAAGAAGGTCTTCGGATTGTA *5845083*

Query 421   AAGCACTTTAAGTTGGGAGGAAGGGTTGTAGATTAATA
CTCTGCAATTTTGACGTTACCG *480*
|||||||||||||||||||||||||||||||||||||||||||||||||||
|||||||||||
Sbjct 5845082   AAGCACTTTAAGTTGGGAGGAAGGGTTGTAGATTAATA
CTCTGCAATTTTGACGTTACCG *5845023*

```
Query 481      ACAGAATAAGCACCGGCTAACTCTGTGCCAGCAGCCG
               CGGTAATACAGAGGGTGCAAGCG 540
               |||||||||||||||||||||||||||||||||||||
               |||||||||
Sbjct 5845022  ACAGAATAAGCACCGGCTAACTCTGTGCCAGCAGCCG
               CGGTAATACAGAGGGTGCAAGCG 5844963

Query 541      TTAATCGGAATTACTGGGCGTAAAGCGCGCGTAGGTGG
               TTCGTTAAGTTGGATGTGAAAT 600
               ||||||||||||||||||||||||||||||||||||||
               |||||||||
Sbjct 5844962  TTAATCGGAATTACTGGGCGTAAAGCGCGCGTAGGTGG
               TTCGTTAAGTTGGATGTGAAAT 5844903

Query 601      CCCCGGGCTCAACCTGGGAACTGCATTCAAAACTGTC
               GAGCTAGAGTATGGTAGAGGGTG 660
               |||||||||||||||||||||||||||||||||||||
               |||||||||
Sbjct 5844902  CCCCGGGCTCAACCTGGGAACTGCATTCAAAACTGTC
               GAGCTAGAGTATGGTAGAGGGTG 5844843

Query 661      GTGGAATTTCCTGTGTAGCGGTGAAATGCGTAGATATA
               GGAAGGAACACCAGTGGCGAAG 720
               ||||||||||||||||||||||||||||||||||||||
               |||||||||
Sbjct 5844842  GTGGAATTTCCTGTGTAGCGGTGAAATGCGTAGATATA
               GGAAGGAACACCAGTGGCGAAG 5844783

Query 721      GCGACCACCTGGACTGATACTGACACTGAGGTGCGAA
               AGCGTGGGGAGCAAACAGGATTA 780
               |||||||||||||||||||||||||||||||||||||
               |||||||||
Sbjct 5844782  GCGACCACCTGGACTGATACTGACACTGAGGTGCGAA
               AGCGTGGGGAGCAAACAGGATTA 5844723

Query 781      GATACCCTGGTAGTCCACGCCGTAAACGATGTCAACTA
               GCCGTTGGGAGCCTTGAGCTCT 840
               ||||||||||||||||||||||||||||||||||||||
               |||||||||
Sbjct 5844722  GATACCCTGGTAGTCCACGCCGTAAACGATGTCAACTA
               GCCGTTGGGAGCCTTGAGCTCT 5844663
```

```
Query 841    TAGTGGCGCAGCTAACGCATTAAGTTGACCGCCTGGG
             GAGTACGGCCGCAAGGTTAAAAC 900
             |||||||||||||||||||||||||||||||||||||
             |||||||||
Sbjct 5844662 TAGTGGCGCAGCTAACGCATTAAGTTGACCGCCTGGG
             GAGTACGGCCGCAAGGTTAAAAC 5844603

Query 901    TCAAATGAATTGACGGGGGCCCGCACAAGCGGTGGAG
             CATGTGGTTTAATTCGAAGCAAC 960
             |||||||||||||||||||||||||||||||||||||
             |||||||||
Sbjct 5844602 TCAAATGAATTGACGGGGGCCCGCACAAGCGGTGGAG
             CATGTGGTTTAATTCGAAGCAAC 5844543

Query 961    GCGAAGAACCTTACCAGGCCTTGACATCCAATGAACT
             TTCCAGAGATGGATTGGTGCCTT 1020
             |||||||||||||||||||||||||||||||||||||
             |||||||||
Sbjct 5844542 GCGAAGAACCTTACCAGGCCTTGACATCCAATGAACT
             TTCCAGAGATGGATTGGTGCCTT 5844483

Query 1021   CGGGAGCATTGAGACAGGTGCTGCATGGCTGTCGTCA
             GCTCGTGTCGTGAGATGTTGGGT 1080
             ||||| |||||||||||||||||||||||||||||||
             |||||||||
Sbjct 5844482 CGGGAACATTGAGACAGGTGCTGCATGGCTGTCGTCA
             GCTCGTGTCGTGAGATGTTGGGT 5844423

Query 1081   TAAGTCCCGTAACGAGCGCAACCCTTGTCCTTAGTTAC
             CAGCACGTTATGGTGGGCACTC 1140
             ||||||||||||||||||||||||||||||||||||||
             |||||||||
Sbjct 5844422 TAAGTCCCGTAACGAGCGCAACCCTTGTCCTTAGTTAC
             CAGCACGTTATGGTGGGCACTC 5844363

Query 1141   TAAGGAGACTGCCGGTGACAAACCGGAGGAAGGTGG
             GGATGACGTCAAGTCATCATGGCC 1200
             |||||||||||||||||||||||||||||||||||||
             |||||||||
Sbjct 5844362 TAAGGAGACTGCCGGTGACAAACCGGAGGAAGGTGG
             GGATGACGTCAAGTCATCATGGCC 5844303
```

Query 1201 CTTACGGCCTGGGCTACACACGTGCTACAATGGTCGGT
ACAAAGGGTTGCCAAGCCGCGA 1260
||||||||||||||||||||||||||||||||||||||||||||||||||
|||||||||

Sbjct 5844302 CTTACGGCCTGGGCTACACACGTGCTACAATGGTCGGT
ACAAAGGGTTGCCAAGCCGCGA 5844243

Query 1261 GGTGGAGCTAATCCCATAAAACCGATCGTAGTCCGGAT
CGCAGTCTGCAACTCGACTGCG 1320
||||||||||||||||||||||||||||||||||||||||||||||||||
|||||||||

Sbjct 5844242 GGTGGAGCTAATCCCATAAAACCGATCGTAGTCCGGAT
CGCAGTCTGCAACTCGACTGCG 5844183

Query 1321 TGAAGTCGGAATCGCTAGTAATCGTGAATCAGAATGTC
ACGGTGAATACGTTCCCGGGCC 1380
||||||||||||||||||||||||||||||||||||||||||||||||||
|||||||||

Sbjct 5844182 TGAAGTCGGAATCGCTAGTAATCGTGAATCAGAATGTC
ACGGTGAATACGTTCCCGGGCC 5844123

Query 1381 TTGTACACACCGCCCGTCACACCATGGGAGTGGGTTG
CACCAGAAGTAGCTAGTCTAACC 1440
||||||||||||||||||||||||||||||||||||||||||||||||||
|||||||||

Sbjct 5844122 TTGTACACACCGCCCGTCACACCATGGGAGTGGGTTG
CACCAGAAGTAGCTAGTCTAACC 5844063

Query 1441 TTCGGGAGGACGGTTACCACGGTGTGATTCATGACTGG
GGTGAAGTCGTAACAAGGTAGC 1500
||||||||||||||||||||||||||||||||||||||||||||||||||
|||||||||

Sbjct 5844062 TTCGGGAGGACGGTTACCACGGTGTGATTCATGACTGG
GGTGAAGTCGTAACAAGGTAGC 5844003

Query 1501 CGTAGGGGAACCTGCGGCTGGATCACCTCCTT 1532
||||||||||||||||||||||||||||||||
Sbjct 5844002 CGTAGGGGAACCTGCGGCTGGATCACCTCCTT 5843971

EZBioCloud

Bacteria and Archaea taxonomy, ecology, genomics, metagenomics, and microbiome are the main topics of CJ Bioscience's public data and analytics platform, EzBioCloud. It replaced old databases, EzTaxon, EzTaxon-e, and EzGenome, with a new cloud service that offers bioinformatics tools. The relatively recent development of technology that sequence DNA has made it possible to make use of genome sequencing data, which in turn provides techniques for the classification and identification of members of the bacterial and archaeal kingdoms that are more informative and exact. Building a genome database that contains accurate taxonomic information is of the utmost importance in order to enhance our efforts in exploring prokaryotic diversity and discovering novel species, as well as for routine identifications. This is due to the fact that the current definition of species is based on the comparison of genome sequences between strains of the same species and other strains of the same species. The taxonomic hierarchy of bacteria and archaea is stored in an integrated database known as EzBioCloud. This hierarchy is represented by genome sequences and 16S rRNA gene sequences that have been subjected to quality control. Whole genome assemblies that were stored in the NCBI Assembly Database were analyzed to determine whether or not they were of poor quality. These assemblies were then put through a composite identification bioinformatics pipeline, which involves gene-based searches followed by the determination of the average nucleotide identity. As a consequence of this, the database contains 61700 species and phylotypes, 13132 of which have names that have been published in a valid manner, and 62362 whole-genome assemblies that were identified taxonomically at the genus, species, and subspecies levels at the time of publication of the original article by Yoon et al., 2017 in the International Journal of Systematic and Evolutionary Microbiology (which was formerly known as the International Journal of Systematic Bacteriology). For each genus or higher taxonomic group, genomic parameters such as genome size and DNA G+C content, as well as the occurrence in human microbiome data, were computed. The genome-based categorization and identification of members of the Bacteria and Archaea have been speeded up because of this unified database of taxonomy, 16S rRNA gene, and genome sequences, which is accompanied by bioinformatics tools. At www.ezbiocloud.net/, one can access the database as well as the search tools that are associated with it (Yoon et al., 2017).

Pictorial representation of steps for identification of bacteria and archaea using 16S rRNA gene sequence

Step-I: Data collection of type strain sequences for making server database ready for identification of any newly isolated bacteria or archaea

The reference 16S rRNA gene sequences were kept up to date in the same manner that was detailed earlier (Kim et al., 2012). Through the utilization of the following technique, we endeavoured to identify a sequence that possessed the highest possible quality for each species. A selection was made for the sequence that was derived from the

Whole-Genome Assembly (WGA) of a type strain in situations when there were several sequences available for that type strain. When it came to sequences that were produced via PCR, the quality of the sequencing was manually assessed by engaging in secondary-structure-aware alignment with the help of the EzEditor tool (Jeon et al., 2014). The RAxML software was used to create maximum-likelihood phylogenetic trees of each taxonomic group, such as phyla, classes, orders, or families, from manually aligned 16S rRNA gene sequences (Stamatakis, 2014). These trees were generated from manually aligned sequences. As part of the comprehensive taxonomic hierarchy, which included phylum, class, order, family, genus, and species (subspecies if applicable) (Yoon et al., 2017), all 16S rRNA gene sequences were assigned taxonomically to the species level. This was done in accordance with the taxonomic classification system.

Step-I: Click on https://www.ezbiocloud.net/identify and login to your authorized account

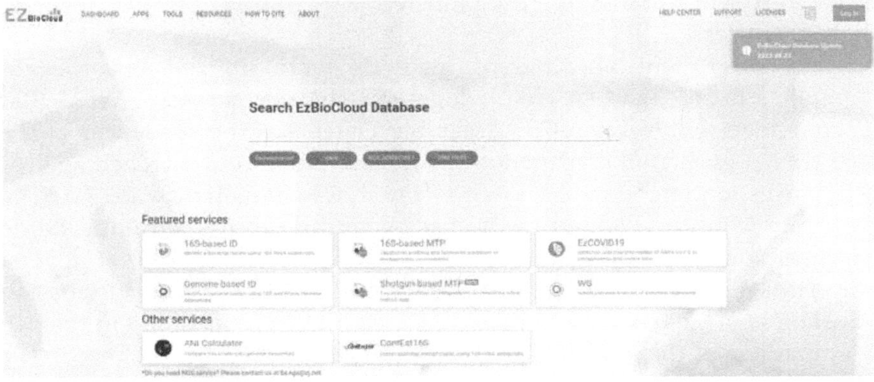

Step-II: Click on identify new sequence and say next...

'Identify' engine

As part of the 'Identify' function, pairwise sequence similarity values between a query sequence and the reference sequences stored in the EzBioCloud database are made available to users. A two-step technique is utilized in order to guarantee that the search engine will locate the sequence that is the most comparable to the one that is contained within the 16S rRNA gene sequence database (Kim et al., 2012). Finding sequences that are comparable is the initial step, followed by the calculation of pairwise sequence similarity values that are useful from a taxonomical perspective (Myers and Miller, 1988). In this instance, EzBioCloud makes use of the usearch software (Edgar, 2010) rather than blastn in order to complete the search process more quickly.

The identification scheme for the sequences of the genome

In order to carry out the taxonomic identification of each WGA, the algorithm that was described in the original research publication written by Yoon et al. in 2017 was utilized. Prior to this, every single WGA was processed by a genome annotation pipeline, which consisted of a collection of software tools and databases. The 16S rRNA gene sequence database, which is also utilized in the 'Identify' engine discussed earlier, and the Reference Genome Database (RefGD) were the two types of databases that were utilized in this study, both of these databases were used. This latter was assembled to hold tetra-nucleotide compositions (Teeling et al., 2004), as well as gyrB and recA sequences from all of the genome sequences of type or representative strains that were available. Through the utilization of an in-house Java software, tetra-nucleotide compositions were determined for each WGA individually. During the time that sequences were being processed in the genome annotation pipeline, the 16S rRNA, gyrB, and recA genes identified in WGAs were predicted. Following that, usearch-based searches were conducted with the purpose of locating RefGD entries.

An assortment of different methods were utilized in order to compile a list of taxa that are phylogenetically related to each WGA that is contained within the NCBI Assembly Database. The sequences of the 16S rRNA, gyrB, and recA genes of a query WGA were searched against the corresponding databases wherever it was possible to do so. The best hits were then added to the list. According to the tetra-nucleotide composition, the correlation values (=z score) were computed against all of the WGAs that were included in our reference group discussion (Teeling et al., 2004). Additionally, the best hits were included in the list. Comparison of the ANI values between the query WGA and those in the list, for which we selected 95% as the limit for species delineation, was the method that was utilized to carry out the final identification. For the purpose of ANI calculation, the OrthoANI method (Lee et al., 2016) was utilized with usearch rather than blastn in order to decrease the amount of time required for computation. An effort attempted to identify all WGAs in terms of their taxonomic classification, at least down to the genus level. In the event that this was not feasible due to the absence of sequences for the 16S rRNA, gyrB, and recA genes, the WGA was designated as "Unidentified" (Yoon et al., 2017).

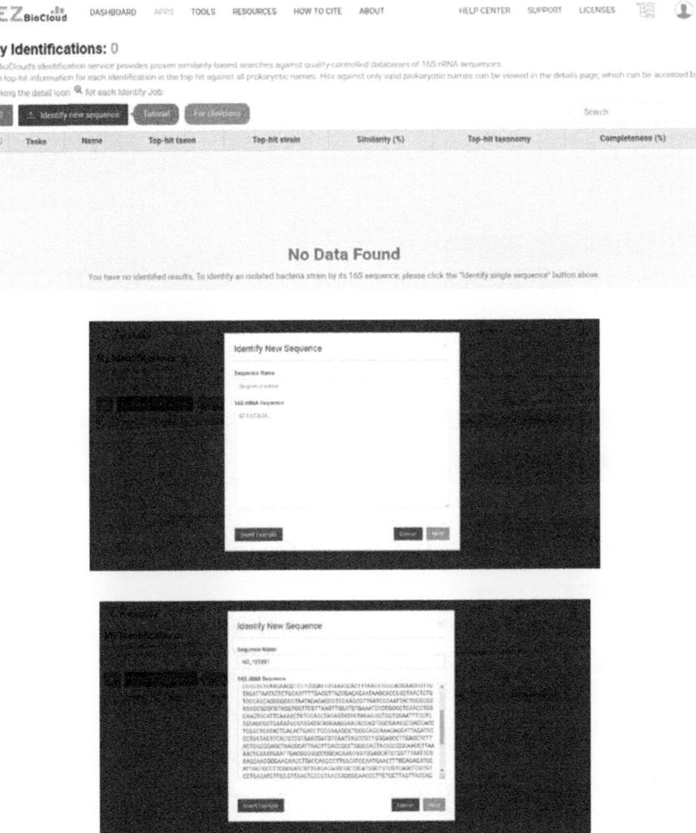

Step-III: Review sequence information and submit

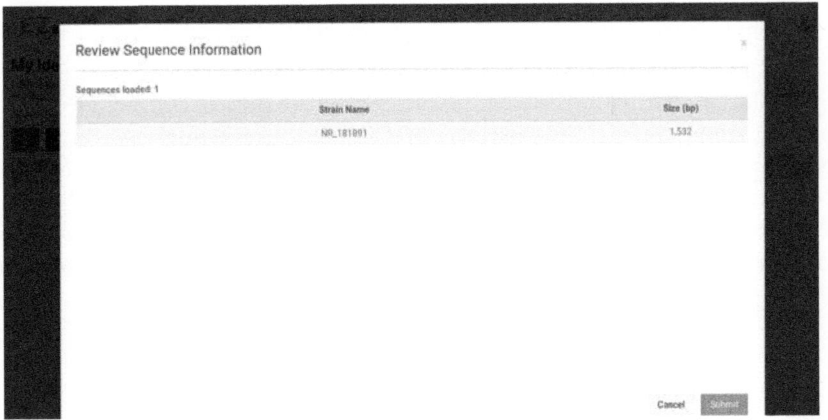

Step-IV: Result will be generated based on similarity-based searches against quality controlled databases (of type-strain (T) sequence data) of 16S rRNA gene sequences.

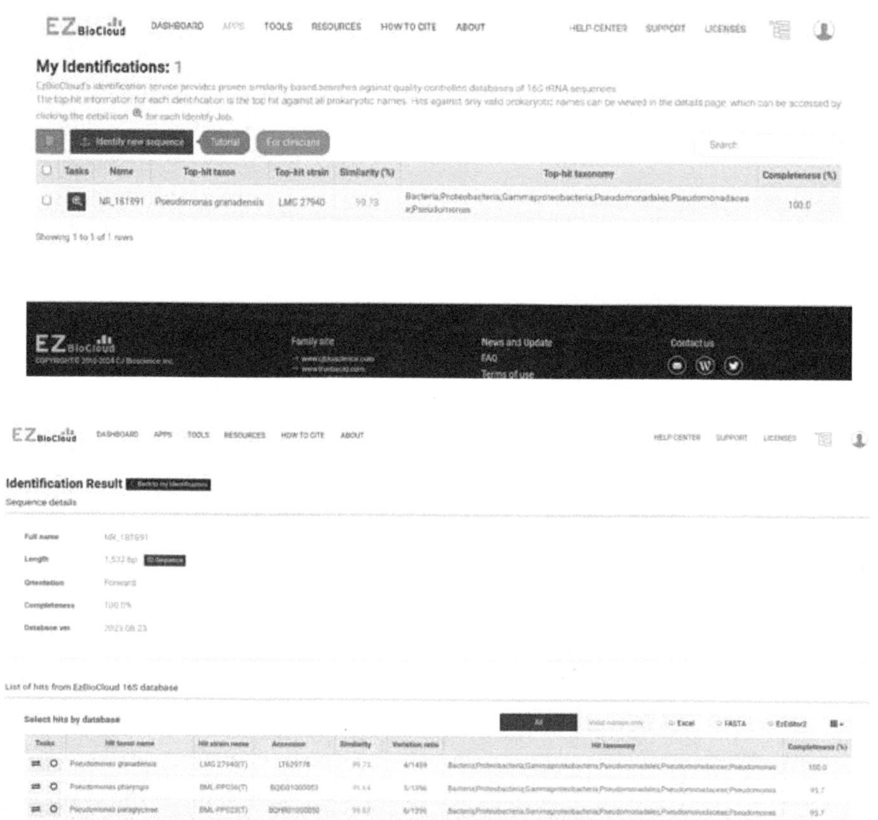

Genomic feature computation for every taxon

A number of taxonomic-relevant genomic parameters were computed and assembled statistically for each taxon, including genome sizes, DNA G+C content, gene counts, and lengths of CDS (coding sequences)/intergenic sections. For all statistical studies, the R software was utilized. Complete genome sequences were used to determine the number of 16S rRNA genes present in each genome. In the event that a species lacked full genomes, Yoon et al. (2017) employed PICRUSt (Langille et al., 2013) to predict the values.

Analysis of the human microbiome's bacterial community

The US NIH human microbiome project's bacterial community dataset was acquired from hmpdacc.org/ and subjected to the bioinformatics workflow. Yoon et al. (2017) gathered and displayed the frequencies of every taxon (ranging from phylum to genera) across 18 body regions of healthy participants as boxplots on the taxon's webpage.

Operating system and software development

The complete system was developed using the Linux operating system and uploaded to servers hosted by Amazon Web Services (AWS). The database management system utilized was MySQL, and the programming languages employed were Java, JavaScript, and R (Yoon et al., 2017).

Acknowledgements

The author duly acknowledges website contents adopted as per CC BY 4.0 from NCBI (https://www.ncbi.nlm.nih.gov/) and EZBioCloud (https://www.ezbiocloud.net/i).

References

Alachiotis, N., Low, T. M. and Pavlidis, P. (2022). Editorial: Scalable bioinformatics: methods, software tools, and hardware architectures. Frontiers in Genetics 12: 822986. https://doi.org/10.3389/fgene.2021.82298.

Altschul, S. F., Gish, W., Miller, W., Myers, E. W. and Lipman, D. J. (1990). Basic local alignment search tool. Journal of Molecular Biology 215(3): 403–410. https://doi.org/10.1016/S0022-2836(05)80360-2.

Camacho, C., Coulouris, G., Avagyan, V., Ma, N., Papadopoulos, J., Bealer, K. et al. (2009). BLAST+: architecture and applications. BMC Bioinformatics 10: 421. https://doi.org/10.1186/1471-2105-10-421.

Connon, S. A. and Giovannoni, S. J. (2002). High-throughput methods for culturing microorganisms in very-low-nutrient media yield diverse new marine isolates. Appl. Environ. Microbiol. 68: 3878–85.

Edgar, R. C. (2010). Search and clustering orders of magnitude faster than BLAST. Bioinformatics (Oxford, England) 26(19): 2460–2461. https://doi.org/10.1093/bioinformatics/btq461.

Hugenholtz, P., Skarshewski, A. and Parks, D. H. (2016). Genome-based microbial taxonomy coming of age. Cold Spring Harb Perspect Biol. 8: a018085.

Jeon, Y. S., Lee, K., Park, S. C., Kim, B. S., Cho, Y. J., Ha, S. M. et al. (2014). EzEditor: a versatile sequence alignment editor for both rRNA- and protein-coding genes. International Journal of Systematic and Evolutionary Microbiology 64(Pt 2): 689–691. https://doi.org/10.1099/ijs.0.059360-0.

Kim, O. S., Cho, Y. J., Lee, K., Yoon, S. H., Kim, M., Na, H. et al. (2012). Introducing EzTaxon-e: a prokaryotic 16S rRNA gene sequence database with phylotypes that represent uncultured species. International Journal of Systematic and Evolutionary Microbiology 62(Pt 3): 716–721. https://doi.org/10.1099/ijs.0.038075-0.

Konstantinidis, K. T. and Tiedje, J. M. (2005). Towards a genome-based taxonomy for prokaryotes. J. Bacteriol. 187: 6258–6264.

Langille, M., Zaneveld, J., Caporaso, J., McDonald, D., Knights, D., Reye, J. A. et al. (2013). Predictive functional profiling of microbial communities using 16S rRNA marker gene sequences. Nat. Biotechnol. 31: 814–821. https://doi.org/10.1038/nbt.2676.

Lee, I., Ouk Kim, Y., Park, S. C. and Chun, J. (2016). OrthoANI: An improved algorithm and software for calculating average nucleotide identity. International Journal of Systematic and Evolutionary Microbiology 66(2): 1100–1103. https://doi.org/10.1099/ijsem.0.000760.

Martiny, A. C. (2019). High proportions of bacteria are culturable across major biomes. ISME J. 13: 2125–2128.

Mount, D. W. (2007). Using the Basic Local Alignment Search Tool (BLAST). CSH Protocols pdb.top17. https://doi.org/10.1101/pdb.top17.

Myers, E. W. and Miller, W. (1988). Optimal alignments in linear space. Computer Applications in the Biosciences : CABIOS 4(1): 11–17. https://doi.org/10.1093/bioinformatics/4.1.11.

NCBI-BLAST. National Center for Biotechnology Information-Basic Local Alignment Search Tool. https://blast.ncbi.nlm.nih.gov/Blast.cgi Asssesed on 21-02-2024.

Rekadwad, B. N., Lian, Z. H., Jiao, J. Y. and Li, W. J. (2023). Chapter 3: Technologies promoting genomic-based taxonomy. Li, W. J., Jiao, J. Y., Salam, N. and Narsing Rao, M. P. (eds.). Modern Taxonomy of Bacteria and Archaea: New Methods, Technology and Advances. Springer Nature. ISBN: 978-981-99-5719-4. https://doi.org/10.1007/978-981-99-5720-0_3.

Romalde, J. L., Balboa, S. and Ventosa, A. (2019). Editorial: microbial taxonomy, phylogeny and biodiversity. Front Microbiol. 10: 1324.

Rosselló-Móra, R. and Whitman, W. B. (2019). Dialogue on the nomenclature and classification of prokaryotes. Syst. Appl. Microbiol. 42: 5–14.

Saeed, U. and Usman, Z. (2019). Biological sequence analysis. *In*: Husi, H. (ed.). Computational Biology. Codon Publications.

Stamatakis, A. (2014). RAxML version 8: a tool for phylogenetic analysis and post-analysis of large phylogenies. Bioinformatics (Oxford, England) 30(9): 1312–1313. https://doi.org/10.1093/bioinformatics/btu033.

Teeling, H., Meyerdierks, A., Bauer, M., Amann, R. and Glöckner, F. O. (2004). Application of tetranucleotide frequencies for the assignment of genomic fragments. Environmental Microbiology 6(9): 938–947. https://doi.org/10.1111/j.1462-2920.2004.00624.x.

Wheeler, D. and Bhagwat, M. (2007). BLAST QuickStart: Example-Driven Web-Based BLAST Tutorial. In: Bergman NH, editor. Comparative Genomics: Volumes 1 and 2. Totowa (NJ): Humana Press; 2007. Chapter 9. Available from: https://www.ncbi.nlm.nih.gov/books/NBK1734/.

Yoon, S. H., Ha, S. M., Kwon, S., Lim, J., Kim, Y., Seo, H. et al. (2017). Introducing EzBioCloud: a taxonomically united database of 16S rRNA gene sequences and whole-genome assemblies. International Journal of Systematic and Evolutionary Microbiology 67(5): 1613–1617. https://doi.org/10.1099/ijsem.0.001755.

Chapter 6
Next-Generation Sequencing (NGS) Workflow
Amplicon Sequencing for Barcoding and Integrative Taxonomy

Introduction

The Next-Generation Sequencing (NGS) has revolutionized the field of molecular biology by enabling the rapid and cost-effective analysis of large-scale genetic data. The field of next-generation sequencing, or NGS, has completely changed how we think about biological systems. Clinical research has made very major and crucial advancements as a result of it. Researchers and medical professionals can examine at the DNA level the mechanisms associated with uncommon genetic diseases, cancer, newborn infections, and other diseases. Thanks to this creative thinking, tailored treatments and diagnostics have been made possible. Researchers can sequence from specific targeted sections to the complete human genome in a single day thanks to NGS platforms, which can simultaneously sequence millions of DNA fragments. Large-scale genomic sequencing has also been made possible by NGS, which is far more advantageous than earlier sequencing methods (Roche; https://rochesequencingstore.com/resources/what-is-next-generation-sequencing/). This advancement has particularly transformed the landscape of DNA barcoding and integrative taxonomy, offering unprecedented opportunities for biodiversity research and species identification (Shokralla et al., 2014; 2015; Taberlet et al., 2012;

Ji et al., 2013). The paper by Cruaud et al. (2017) presents a comprehensive exploration of NGS technologies applied to these domains, emphasizing the efficacy of sequencing multiple amplicons for barcoding purposes and its implications for integrative taxonomy (Shokralla et al., 2015; Meier et al., 2016).

Traditionally, DNA barcoding has relied on sequencing a single genetic marker, such as the mitochondrial cytochrome c oxidase subunit 1 (COI) gene, to distinguish between species. While effective for many taxa, this approach may encounter limitations in resolving closely related species or populations with recent divergence. NGS techniques address these challenges by allowing the simultaneous sequencing of multiple genomic regions across numerous samples, thereby enhancing both the resolution and accuracy of species identification (Meyer and Paulay, 2005; Bergsten et al., 2012). The integration of NGS data with traditional taxonomic methods forms the basis of integrative taxonomy, which combines molecular, morphological, ecological, and behavioural data to provide a more holistic understanding of species boundaries and evolutionary relationships. By incorporating a diverse array of genetic markers, NGS facilitates the elucidation of species complexes, cryptic diversity, and evolutionary processes that shape biodiversity patterns. In their study, Cruaud et al. (2017) showcase the utility of NGS for barcoding and integrative taxonomy through a detailed examination of various amplicon sequencing approaches. They highlight the advantages of targeting multiple genomic regions, such as the ribosomal RNA gene cluster (rDNA) and other mitochondrial and nuclear markers, in resolving taxonomic uncertainties and improving species delimitation. Furthermore, the authors demonstrate the scalability and efficiency of NGS methodologies, paving the way for large-scale biodiversity assessments and ecological monitoring. Hence, the application of NGS in DNA barcoding and integrative taxonomy represents a transformative paradigm shift in biodiversity research, offering unprecedented insights into species diversity, evolutionary processes, and conservation priorities. The paper by Cruaud et al. (2017) serves as a valuable contribution to this rapidly evolving field, elucidating the potential of NGS technologies to revolutionize our understanding of the natural world (Berry et al., 2011; Seitz et al., 2015; Levy et al., 2015; Herbold et al., 2015; McClenaghan et al., 2015; Gibson et al., 2015; Fahner et al., 2016; Cruaud et al., 2017).

Advantages of NGS technology

Sanger sequencing, also known as first-generation sequencing, had been widely utilized over the past three decades, which resulted in substantial advancements in the understanding of the human genome. In spite of this, next-generation sequencing (NGS) has eclipsed Sanger sequencing due to the numerous benefits that NGS offers. A number of advantages are associated with next-generation sequencing (NGS), including enhanced sensitivity and coverage, cost-effectiveness, and efficient workflow, as well as a reduction in turnaround time (Behjati and Tarpey, 2013; Schuster, 2008).

Types of NGS applications

Depending on area or research and interest of researchers, various companies have developed sequencing platforms (Table 1a and 1b). Following are some applications of NGS technology:

- *Whole-Genome Sequencing (WGS)*: Whole-genome sequencing, also known as WGS, is an all-encompassing technique for figuring out the contents of complete genomes. Identifying inherited illnesses, describing the mutations that drive the growth of cancer, and monitoring disease outbreaks have all been made significantly easier with the help of genomic information. Whole-genome sequencing is a significant tool for genomics research since these days, sequencing prices are falling at a rapid rate, and modern sequencers are able to generate vast amounts of data (Source: Illumina Inc.).

- *Whole-Exome Sequencing (WES)*: A method of next-generation sequencing (NGS) that involves sequencing the portions of the genome that code for proteins is known as whole-exome sequencing. This method is frequently utilized. This method is a cost-effective alternative to whole-genome sequencing because the human exome, which accounts for less than 2% of the genome, contains around 85% of the known variations that are associated with disease (van Dijk et al., 2014). It is possible to quickly detect coding variants across a wide variety of applications by using exome sequencing in conjunction with exome enrichment. These applications include population genetics, genetic disease research, and cancer research primarily.

- *Targeted Sequencing (TS)*: The analysis of specific mutations in a given sample can be simplified with the help of targeted gene sequencing panels, which are important tools. In focused panels, a specific group of genes or gene regions are included. These genes or regions are recognized to have known or suspected correlations with the disease or trait that is being investigated. Gene panels can be purchased with content that has already been picked, or they can be built specifically to incorporate sequences of interest in the genome (Source: Illumina Inc.). The clinical application of a great number of well-known gene alterations that are responsible for the pathogenesis of disease, such as cancer driver genes, has been widespread. For the purposes of diagnosis, prognosis, therapy monitoring, and other purposes, TS panels concentrate their attention on a particular number of these genes. As a result, the utilization of TS panels in clinical settings will result in a reduction in costs, as well as an increase in confidence and improved chances for insurance reimbursement. TS has a higher sequencing depth of coverage (1000× or higher) than non-NGS-based methods like ARMS, PCR, AS-PCR, and BEAMing. This is because it is based on a special technique called allele-specific amplification refractory mutation system. Using this method, it is possible to identify mutations that are only found in a tiny portion of malignant cells. Furthermore, it is feasible to detect a Variation Allele Frequency (VAF) as low as 0.1–0.2% when it comes to detecting minimal residual disease. Furthermore, due to the fact that mutations that result in truncation or potential mRNA attenuation in any region of the tumour suppressor genes can be regarded clinically significant, it is not possible for the technologies that were discussed above to detect the entire sections of genes that are associated to tumours (Pei et al., 2023).

- *Epigenomics*: Epigenomics is the comprehensive examination of epigenetic modifications in many genes inside a cell or organism. Analysis of both epigenetics and epigenomics may include the examination of changes in DNA methylation, DNA-protein interactions, chromatin accessibility, histone modifications, and other related factors (Source: Illumina Inc.; Gomase and Tagore, 2008; Agarwal et al., 2020).

- *RNA Sequencing (RNA-Seq)*: RNA-Seq is a modern method for analyzing the transcriptome that utilizes advanced deep-sequencing methods. Research conducted with this approach has already changed our understanding of the size and intricacy of eukaryotic transcriptomes. RNA-Seq offers a more accurate assessment of transcript levels and their isoforms compared to alternative approaches (Koch et al., 2018; Wang et al., 2009; Zhang et al., 2018).

- *PCR Techniques in Next-Generation Sequencing (PCR-NGS or NGS-PCR)*: NGS has been widely used in the clinical setting, thus it would seem that methods like PCR would not be in high demand. This is incorrect, though, as PCR methods are crucial to the development of NGS technology. Although whole genome sequencing is still challenging to implement in a clinical laboratory due to high sequencing costs and problems with data processing, analysis, and storage, which can decrease efficiency and increase turnaround times, NGS has quickly grown in importance as a component of clinical molecular diagnostics. As targeted sequencing is more efficient, it is therefore frequently utilized in clinical diagnosis. Targeted NGS sequencing relies heavily

Table 1a. Various companies provides platforms for next-generation sequencing.

Company	Technology	Key Features	Applications
Illumina	Sequencing by Synthesis (SBS)	High throughput, short-read sequencing, market dominance	Research, diagnostics, personalized medicine, forensics
Thermo Fisher Scientific (Ion Torrent)	Ion Semiconductor Sequencing	Fast turnaround time, lower cost per run (for smaller projects)	Targeted sequencing, microbial sequencing
Pacific Biosciences (PacBio)	Single Molecule Real-Time Sequencing (SMRT)	Long-read sequencing, high accuracy for complex regions	Structural variant detection, genome assembly, isoform sequencing
Oxford Nanopore Technologies	Nanopore Sequencing	Long-read sequencing, real-time analysis, portability	Field sequencing, rapid diagnostics, direct RNA sequencing
BGI Genomics	Combinatorial Probe-Anchor Synthesis (cPAS, based on DNBseq technology)	High throughput, competitive pricing	Large-scale sequencing projects, population genomics
MGI Tech	The MGISeq platforms utilize a unique sequencing technology called DNBSeq™. This involves creating DNA Nanoballs (DNBs) through rolling circle replication, which are then loaded onto patterned flow cells for sequencing.	Competitive ranges of sequencers	High-medium-low-throughput sequencing applications

Table 1b. Various companies provides platforms for next-generation sequencing.

Company	Sequencer(s)	Technology	Key Features
Illumina	AutoLoader, HiScan System, HiSeq 3000 System, HiSeq 4000 System, iScan System, iSeq 100 System, MiniSeq System, MiSeq System, MiSeqDx in Research Mode, NextSeq 1000 System, NextSeq 2000 System, NextSeq 500 System, NextSeq 550 System, NextSeq 550Dx in Research Mode, NovaSeq 6000 System, NovaSeq 6000Dx in Research Mode	Sequencing by Synthesis (SBS)	High throughput (NovaSeq), flexibility and mid-range options (NextSeq), benchtop accessibility (iSeq, MiSeq)
Thermo Fisher Scientific	Ion Torrent Genexus System, SeqStudio Genetic Analyzer, SeqStudio Series Genetic Analyzers for Sanger sequencing, SeqStudio Flex genetic analyzer, SeqStudio Flex Series Genetic Analyzers for Sanger sequencing	Semiconductor sequencing	Fast turnaround time, integrated workflow (Genexus), affordability (S5)
Oxford Nanopore	MinION Mk1B, MinION Compute, MinION Mk1D, GridION, GridION Q-Line, PromethION 2 Solo, PromethION 2 Integrated, PromethION 24, PromethION 48	Nanopore sequencing	Long reads for structural variation, portability (MinION, Flongle)
PacBio	Onso system (Short-read sequencing), Revio system (Long-read sequencing), Sequel IIe system (Long-read sequencing)	Single-Molecule Real-Time (SMRT) sequencing	Very long reads (ideal for complex genomes, structural analysis), high accuracy
MGI Tech	DNBSEQ-T20×2, DNBSEQ-T10×4RS, DNBSEQ-T7, DNBSEQ-T7 for HotMPS Only, DNBSEQ-G400, DNBSEQ-G400* for HotMPS only, DNBSEQ-G50, DNBSEQ-G99	DNA Nanoball (DNB) sequencing	High throughput, affordability

on PCR techniques, which enable the simultaneous sequencing of several targeted regions and the creation of numerous NGS libraries (He et al., 2022; Cheng et al., 2019; Zhou et al., 2021; Goswami, 2016).

Analyses of the MiSeq data set (Figure 1)

Step 1, from read filtering to clustering

Raw sequence data underwent quality control checks using FastQC v.0.11.2 (Figure 2). The paired-end reads were reassembled using FLASH v.1.2.11 with default settings and an extended maximum overlap length of 300. When the paired-end reads did not overlap (COI-long and EF1a, Table 1 and Table 2), the forward and reverse reads were analyzed individually. We utilized CUTADAPT v.1.2.1 with its default settings to organize paired reads by gene region and eliminate primers. Trimming of COI-long and EF1a forward and reverse reads was conducted by quality trimming the reads at the first position with a quality score < 21 using VSEARCH v.1.8.1, which can be found at https://github.com/torognes/vsearch. Following the removal of primers and quality filtering, the fastq files underwent conversion to fasta files. Subsequently, sequences shorter than 150 bp were eliminated utilizing VSEARCH. After the sequences were processed, potential chimeric sequences were eliminated using

NGS: Illumina platform

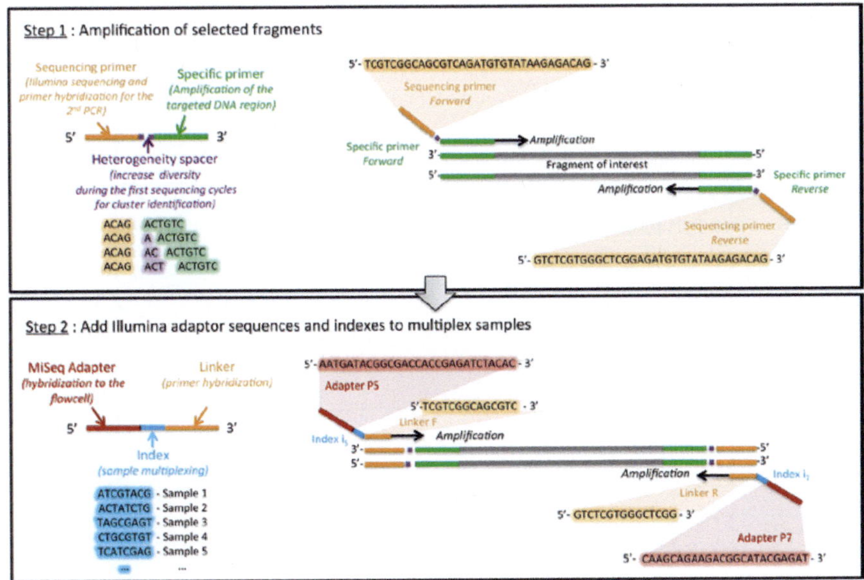

Figure 1. Illustration of the two-step PCR approach (Adopted from Craud et al., 2017).

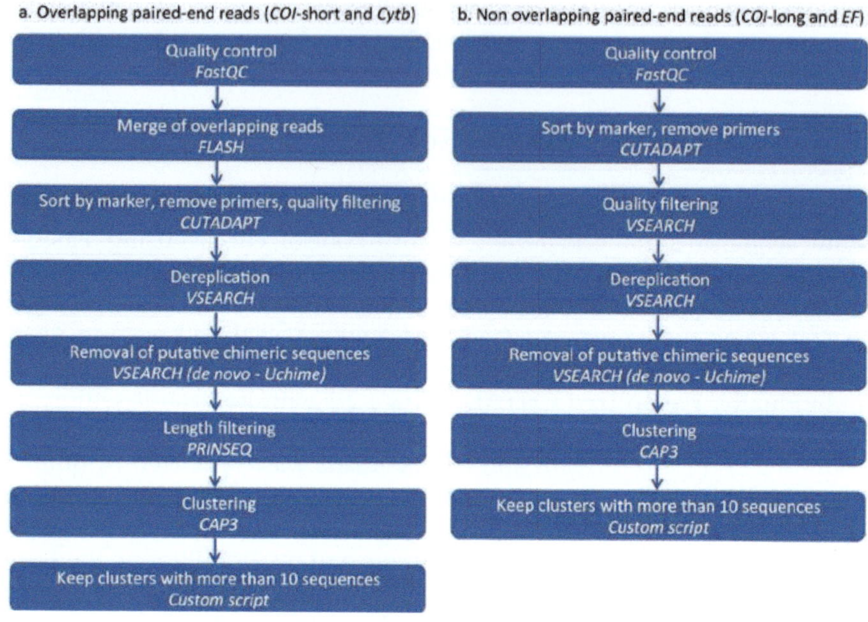

Figure 2. Step 1, from read filtering to clustering.

Table 2. Specifically targeted areas of DNA for amplification (Adopted from Craud et al., 2017).

Gene region	COI			Cytb	EF
Primer pair	LCO1490puc-HCO2198puc	UEA3-HCO2198		CB1-CB2	F2-557F-F2-1118R
	"COI-long"	"COI-short"			
Primer position	LCO1490pucc →	UEA3 →	HCO2198 HCO2198puc ←		
		256	709		
Amplicon size (nt)	658	409		433	518
MiSeq sequenced product (nt) *	709	460		485	563
Read overlap?	no	yes		yes	no

VSEARCH. Trimming of Cytb and COI-short sequences was conducted based on length criteria using PRINSEQ v.0.20.4, with a minimum length of 400 bp and a maximum length of 550 bp. Sequences obtained from Illumina were clustered using SWARM v.2.1.6 and CAP3 with their default settings. Clusters with fewer than 10 sequences were removed from the data sets using VSEARCH. Variations in read lengths resulting from quality trimming caused SWARM to overestimate the cluster count for COI-long and EF1a forward and reverse reads. Thus, only the outcomes achieved using CAP3 were further examined.

Step 2, quality control of clusters of reads

Aligning the consensus sequence of each gene region cluster with the corresponding Sanger data set was done using MAFFT v7.222 with default parameters (Figure 3). When the paired-end reads did not overlap

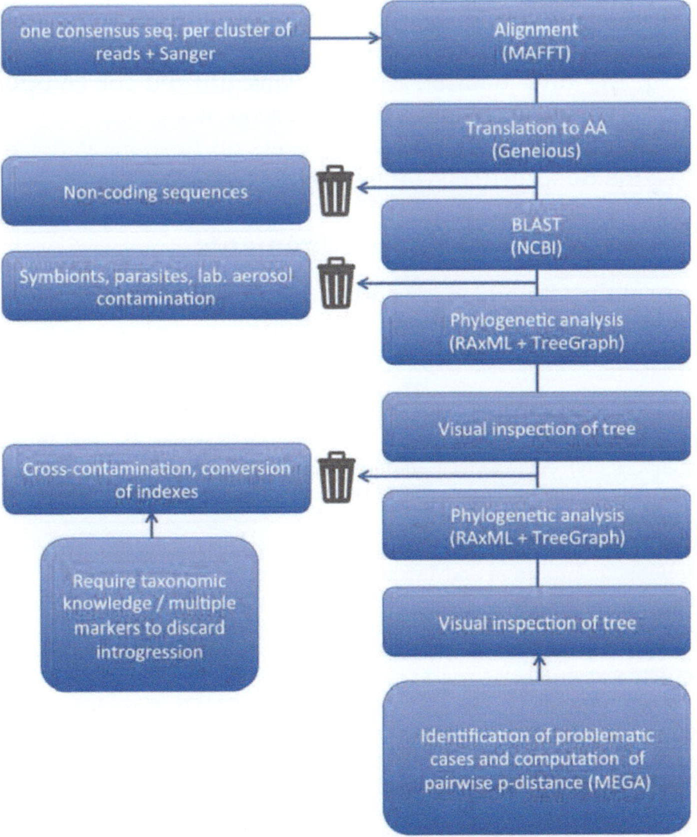

Figure 3. Step 2, quality control of clusters of reads.

(COI-long and EF1a), the analysis was conducted separately for clusters of reads 1 and clusters of reads 2. During this stage, a total of 6 data sets were compiled, including combinations of different sequencing methods for COI, Cytb, and EF genes. Translating alignments into amino acids was done using Geneiousv6.1.6 in order to identify frameshift mutations and premature stop codons. Excluding non-coding sequences from the dataset enhanced the quality of the data. Identifying sequences that did not belong to the target group involved using NCBI-BLAST. Phylogenetic trees were subsequently deduced for each gene region utilizing RAxML. The trees were displayed and labelled using TreeGraph 2. An examination of the trees was conducted to detect any contaminations, which were then eliminated from the data sets. Combining reads 1 and 2 for COI and EF involved merging them into a unified data set with gaps inserted between the non-overlapping reads. After merging the COI-long and COI-short data sets, Geneious v6.1.6 was used. When the COI-long and COI-short sequences matched exactly in their overlapping section, they were combined to create the longest COI sequence possible. If discrepancies were identified, they were not combined, and both sequences were included in the final dataset. MEGA7 was utilized for computing the average divergence (p-distance) between sequence groups in challenging scenarios. You can access the final data sets on figshare at the following link: https://dx.doi.org/10.6084/m9.figshare.4555492.

Acknowledgements

The author duly acknowledges content, table and figures adopted as per CC BY 4.0 from Cruaud et al., 2017; Sci Rep.

References

Agarwal, G., Kudapa, H., Ramalingam, A., Choudhary, D., Sinha, P., Garg, V. et al. (2020). Epigenetics and epigenomics: underlying mechanisms, relevance, and implications in crop improvement. Functional & Integrative Genomics 20(6): 739–761. https://doi.org/10.1007/s10142-020-00756-7.

Behjati, S. and Tarpey, P. S. (2013). What is next generation sequencing?. Archives of disease in childhood. Education and Practice Edition 98(6): 236–238. https://doi.org/10.1136/archdischild-2013-304340.

Bergsten, J., Bilton, D. T., Fujisawa, T., Elliott, M., Monaghan, M. T., Balke, M. et al. (2012). The effect of geographical scale of sampling on DNA barcoding. Systematic Biology 61(5): 851–869. https://doi.org/10.1093/sysbio/sys037.

Berry, D., Ben Mahfoudh, K., Wagner, M. and Loy, A. (2011). Barcoded primers used in multiplex amplicon pyrosequencing bias amplification. Applied and Environmental Microbiology 77(21): 7846–7849. https://doi.org/10.1128/AEM.05220-11.

Cheng, Y. W., Stefaniuk, C. and Jakubowski, M. A. (2019). Real-time PCR and targeted next-generation sequencing in the detection of low level EGFR mutations: Instructive case analyses. Respiratory Medicine Case Reports 28: 100901. https://doi.org/10.1016/j.rmcr.2019.100901.

Cruaud, P., Rasplus, J. Y., Rodriguez, L. J. and Cruaud, A. (2017). High-throughput sequencing of multiple amplicons for barcoding and integrative taxonomy. Scientific Reports 7: 41948. https://doi.org/10.1038/srep41948.

Fahner, N. A., Shokralla, S., Baird, D. J. and Hajibabaei, M. (2016). Large-scale monitoring of plants through environmental DNA metabarcoding of soil: recovery, resolution, and annotation of four DNA markers. PloS One 11(6): e0157505. https://doi.org/10.1371/journal.pone.0157505.

Fazzari, M. J. and Greally, J. M. (2010). Introduction to epigenomics and epigenome-wide analysis. Methods in Molecular Biology (Clifton, N.J.) 620: 243–265. https://doi.org/10.1007/978-1-60761-580-4_7.

Gibson, J. F., Shokralla, S., Curry, C., Baird, D. J., Monk, W. A., King, I. et al. (2015). Large-scale biomonitoring of remote and threatened ecosystems via high-throughput sequencing. PloS One 10(10): e0138432. https://doi.org/10.1371/journal.pone.0138432.

Gomase, V. S. and Tagore, S. (2008). Epigenomics. Current Drug Metabolism 9(3): 232–237. https://doi.org/10.2174/138920008783884821.

Goswami, R. S. (2016). PCR techniques in next-generation sequencing. Methods in Molecular Biology (Clifton, N.J.) 1392: 143–151. https://doi.org/10.1007/978-1-4939-3360-0_13.

He, C., Wei, C., Wen, J., Chen, S., Chen, L., Wu, Y. et al. (2022). Comprehensive analysis of NGS and ARMS-PCR for detecting EGFR mutations based on 4467 cases of NSCLC patients. Journal of Cancer Research and Clinical Oncology 148(2): 321–330. https://doi.org/10.1007/s00432-021-03818-w.

Herbold, C. W., Pelikan, C., Kuzyk, O., Hausmann, B., Angel, R., Berry, D. et al. (2015). A flexible and economical barcoding approach for highly multiplexed amplicon sequencing of diverse target genes. Frontiers in Microbiology 6: 731. https://doi.org/10.3389/fmicb.2015.00731.

Ji, Y., Ashton, L., Pedley, S. M., Edwards, D. P., Tang, Y., Nakamura, A. et al. (2013). Reliable, verifiable and efficient monitoring of biodiversity via metabarcoding. Ecology Letters 16(10): 1245–1257. https://doi.org/10.1111/ele.12162.

Koch, C. M., Chiu, S. F., Akbarpour, M., Bharat, A., Ridge, K. M., Bartom, E. T. et al. (2018). A Beginner's Guide to Analysis of RNA sequencing data. American Journal of Respiratory Cell and Molecular Biology 59(2): 145–157. https://doi.org/10.1165/rcmb.2017-0430TR.

Levy, S. F., Blundell, J. R., Venkataram, S., Petrov, D. A., Fisher, D. S. and Sherlock, G. (2015). Quantitative evolutionary dynamics using high-resolution lineage tracking. Nature 519(7542): 181–186. https://doi.org/10.1038/nature14279.

McClenaghan, B., Gibson, J. F., Shokralla, S. and Hajibabaei, M. (2015). Discrimination of grasshopper (Orthoptera: Acrididae) diet and niche overlap using next-generation sequencing of gut contents. Ecology and Evolution 5(15): 3046–3055. https://doi.org/10.1002/ece3.1585.

Meier, R., Wong, W., Srivathsan, A. and Foo, M. (2016). $1 DNA barcodes for reconstructing complex phenomes and finding rare species in specimen-rich samples. Cladistics : the International Journal of the Willi Hennig Society 32(1): 100–110. https://doi.org/10.1111/cla.12115.

Meyer, C. P. and Paulay, G. (2005). DNA barcoding: Error rates based on comprehensive sampling. Plos Biology 3: 2229–2238.

Pei, X. M., Yeung, M. H. Y., Wong, A. N. N., Tsang, H. F., Yu, A. C. S., Yim, A. K. Y. et al. (2023). Targeted sequencing approach and its clinical applications for the molecular diagnosis of human diseases. Cells 12(3): 493. https://doi.org/10.3390/cells12030493.

Schuster, S. C. (2008). Next-generation sequencing transforms today's biology. Nature Methods 5(1): 16–18. https://doi.org/10.1038/nmeth1156.

Seitz, V., Schaper, S., Dröge, A., Lenze, D., Hummel, M. and Hennig, S. (2015). A new method to prevent carry-over contaminations in two-step PCR NGS library preparations. Nucleic Acids Research 43(20): e135. https://doi.org/10.1093/nar/gkv694.

Shokralla, S., Gibson, J. F., Nikbakht, H., Janzen, D. H., Hallwachs, W. and Hajibabaei, M. (2014). Next-generation DNA barcoding: using next-generation sequencing to enhance and accelerate DNA barcode capture from single specimens. Molecular Ecology Resources 14(5): 892–901. https://doi.org/10.1111/1755-0998.12236.

Shokralla, S., Porter, T. M., Gibson, J. F., Dobosz, R., Janzen, D. H., Hallwachs, W. et al. (2015). Massively parallel multiplex DNA sequencing for specimen identification using an Illumina MiSeq platform. Scientific Reports 5: 9687. https://doi.org/10.1038/srep09687.

Taberlet, P., Coissac, E., Pompanon, F., Brochmann, C. and Willerslev, E. (2012). Towards next-generation biodiversity assessment using DNA metabarcoding. Molecular Ecology 21(8): 2045–2050. https://doi.org/10.1111/j.1365-294X.2012.05470.x.

van Dijk, E. L., Auger, H., Jaszczyszyn, Y. and Thermes, C. (2014). Ten years of next-generation sequencing technology. Trends in Genetics : TIG 30(9): 418–426. https://doi.org/10.1016/j.tig.2014.07.001.

Wang, Z., Gerstein, M. and Snyder, M. (2009). RNA-Seq: a revolutionary tool for transcriptomics. Nature reviews. Genetics 10(1): 57–63. https://doi.org/10.1038/nrg2484.

Zhang, H., He, L. and Cai, L. (2018). Transcriptome Sequencing: RNA-Seq. Methods in Molecular Biology (Clifton, N.J.) 1754: 15–27. https://doi.org/10.1007/978-1-4939-7717-8_2.

Zhou, X., Chen, X., Jiang, Y., Qi, Q., Hao, N., Liu, C. et al. (2021). A rapid PCR-Free next-generation sequencing method for the detection of copy number variations in prenatal samples. Life (Basel, Switzerland) 11(2): 98. https://doi.org/10.3390/life11020098.

Chapter 7

New Tools for Taxonomic Classification of Uncultivated Microbes and the Seqcode to Classify Them

Introduction

Currently, the advancement of new methodologies helps in extracting environmental DNA (eDNA) or genomic material from any sample, and sophisticated analysis technologies allow us to anticipate and describe previously unseen microbes (Murray and Schleifer, 1994). Yet-to-culture is a concept that restricts physical experiments and descriptions to live culture, while genomic insights can reveal information across all species, whether they are cultivable or uncultivated. A comprehensive description, such as axenic culture, has not been addressed by the ICNP, and the absence of universal standards has restricted the cataloguing of Candidatus. There is a significant issue with the naming of Candidatus taxa in nature and their classification according to the Prokaryotic Code. Thus, Candidatus taxa will have a distinct nomenclatural system, adhering to the principles used for cultured Bacteria and Archaea but with their own set of officially recognized names. Should this system become more prevalent, it could simplify the categorization of the majority of species and create a unified database of legally recognized names, reducing confusion and duplicate entries (Konstantinidis et al., 2017). This action could potentially be a violation of Rule 18f.

New bacteria must be added to at least two foreign culture collections (Rule 30 (3)(c)) before they can be published in the foreign Journal of

Systematic and Evolutionary Microbiology (IJSEM). If a new species name is published somewhere other than IJSEM, the writers should send IJSEM a cover letter along with the published article(s). The standards for validation are the same as those for publishing in IJSEM. For twenty years now, these rules have been in place. There may be a big change in nomenclature in near future. The Prokaryote Code sets the rules for the old method of prokaryotic nomenclature, which is based on pure cultures in the lab (about 20,000 species). The ICSP is against any "official" prokaryotic taxonomy, and the Prokaryotic Code only covers naming that involves any outside microbiological groups getting involved. Scientists may have to pick between the ICSP rules and diversity, which could lead to chaos and the end of a system for names that was already well-organized (Oren and Garrity, 2018).

New tools and rules for nomenclature and taxonomy of prokaryotes

Microbial Genomes Atlas (MiGA) webserver

For the past thirty years, the 16S RNA gene sequence has been utilized as the marker gene for categorizing and differentiating isolates at the species level, however it may not be sufficient to differentiate some genera. Therefore, to improve accuracy in identifying species and more specific levels like subspecies, strain, and pathovar, researchers suggest using Average Nucleotide Identity (ANI) based on whole-genome sequencing (WGS) as a reliable proxy (Rodriguez-R et al., 2018a). The MiGA web server utilizes 16S rRNA and other significant proxies like ANI, AAI, ACG, SAGs, and MAGs to compare unidentified genomic sequences against existing sequences and genomes. In addition, many databases like NCBI Prokaryotic RefSeq Genomes (NCBI Proka RefG) include ACG, SAGs, and MAGs, known as the Clade projects. These are utilized in MiGA to enhance genetic diversity analysis, study inter- and intra-evolutionary relationships among taxa, and identify core genome sets and pangenome (Rodriguez-R et al., 2018b).

The Genome Taxonomy Database (GTDB)

Genomes obtained from pure cultures of Bacteria and Archaea (Mukherjee et al., 2017), along with SAGs and MAGs (Parks et al., 2017; Alneberg et al., 2018), significantly enhanced the overall genomic diversity. The Genome Taxonomy Database (GTDB) provides a taxonomy system based on genomes, with rank-normalized classifications for more than 150,000 bacterial and archaeal genomes from the domain to genus, ensuring phylogenetic consistency. Approximately 40% of the genomes in the

GTDB lack a species identification. To address this limitation, the study utilized the widely recognized Average Nucleotide Identity (ANI) criteria to establish species boundaries and identify species clusters encompassing all available bacterial and archaeal genomes. In contrast to prior ANI investigations, a single representative genome has been selected as the definitive nomenclatural 'type' for each species. Eight thousand seven hundred and ninety two out of 24,706 suggested species clusters are supported by published names. Placeholder names were assigned to the remaining 15,914 species clusters to label the increasing amount of genomes from uncultivated species. This database offers an extensive taxonomic framework from the domain to species level for bacterial and archaeal genomes, aiding research on uncultivated species and enhancing scientific communication (Parks et al., 2020).

The Great Automatic Nomenclator (GAN)

Bacteria and Archaea require new names to accurately reflect their taxonomic position during a period of significant discovery in culturomics, genomics, and metagenomics. Expertise in classical Latin and Greek, as well as awareness of the rules of the ICNP, are required for the time-consuming task of creating taxonomic names. Generating several names required in the multi-omics age is easily accessible. A novel method involves mass-producing names before linking them to taxa, combining Latin and Greek roots with relevant meanings, and using computerized automation for name creation (Pallen, 2021).

The Seqcode

There are three distinct components of taxonomy that come into existence for the purpose of naming germs. These components are classification, nomenclature, and identification. Nomenclature is the only one of these three that is governed by rules, but categorization and identification are not subject to any rules governing their implementation. For the purpose of conveying information about the characteristics of organisms, naming microbes ought to be accompanied with a sequential code (Cowan, 1965). The labels that were obtained for uncultivated prokaryotes belonging to the Candidatus taxonomic group, such as CPR1, CPR3, and SM2F11 (Hug et al., 2016), reveal no traits that are peculiar to any particular species. There are no priority names assigned to the Candidatus taxa because they are cultivated species. Consequently, it is of utmost importance for all species that are found in nature to have a name that is both exact and consistent for the purpose of scientific communication. Despite the fact that the utilization of ribosomal databases, ANI/AAI aggregates, and the utilization of isolates whole-genome provides appropriate species names that are derived

Table 1. Requirements and recommendations for describing cultivated taxa to be named under the ICNP rules. Note that SAM distinguishes between single and multiple strain descriptions, which are treated differently (Permission for use of original content has been obtained by Dr. Bhagwan Rekadwad via CCC RightsLink; License Number: 5742870228563 dated Mar 06, 2024).

Description of cultivated taxa and naming under the ICNP rules
Name and designated type strain
Required
The taxon name and the designated type strain collection numbers need to be given in the main manuscript in the form of a protologue table, and a template is given here. Taxon names need to be written in italics, and the designated types and culture collection numbers should be suffixed with a superscriptT. For example: Salinibacter altiplanensis (AN15T = IBRC-M 11031T = CECT 9105T). The deposit certificates need to be submitted together with the manuscript.
Protologue
Required
If taxa are named under the ICNP rules, protologues must be given in the form of a table in the manuscript. Descriptions in the supplementary material are not allowed because only permanent records are considered for validation of names. For instructions on how to name under the SeqCode rules see Table 2. The required fields for the protologue are available in the online template. The required basic information is: • Name (formed with mnemonic cues). • Etymology. • Designated type strain and two culture collection numbers. • Genome with an INSDC accession number of the designated type strain that meets minimum quality requirements (Bowers et al., 2017). The requirements given in (Table 2) apply to the quality of the genomes. • The 16S rRNA gene accession number of the designated type strain. The gene must be almost complete (> 1,400 bp) if resulting from PCR amplification or complete if resulting from genome sequencing. • In the case of a new genus, designate the type species. • In the case of higher taxa, designate the type genus. • Diagnostic genomic and phenotypic properties, inferred and demonstrated physiological traits. Ideally, a phenotypic discriminative profile using relevant physiological studies is desired.
Optional
The following are optional but highly recommended for more informative taxa descriptions: • Include as much metadata as possible (Field et al., 2008). • As commercial biochemical tests and simple phenotyping using standard laboratory tests often do not reveal relevant properties, a more exhaustive, genome-based description of (potential) phenotypic properties is recommended instead (e.g., metabolic prediction based on functional gene annotation bioinformatics). • It will benefit the study to include ecological information, such as habitat, environmental physicochemical properties, inferred interactions with other organisms, such as symbiosis, syntrophy, and other properties that may be relevant for the description. • Chemotaxonomic properties, such as fatty acids, polar lipids, respiratory quinones or polyamines, unless required by the respective taxonomic subcommittees, may be regarded as optional, given that these can also be determined retrospectively (Sutcliffe et al., 2021). Some fields of relevant metadata are already given in the available online template in order to guide the acquisition of such data, and these can be added as new lines in the protologue template.

Table 1 contd. ...

...Table 1 contd.

Description of cultivated taxa and naming under the ICNP rules
Type material data quality
Required   For taxonomic descriptions with nomenclature according to the ICNP rules, the type material is a living pure culture. This must be deposited in two culture collections from different countries, and the cultures must be fully available to third parties without any restriction, following the recommendation of the ICNP.
Multiple strains ≥ 2   SAM recommends that authors acquire multiple non-clonal strains of the same taxon, ideally from different origins.
Required   With the description of multiple strains, it is required that: • There is proof of non-clonality of the strains. For this purpose, different genome-based methods can be applied. However, for a fast-check at the initial editorial desk-based evaluation, genome fingerprinting techniques, such as PCR-RAPD profiles (Sikorski et al., 1999), should be applied, and an image should be provided in the supplementary material showing distinct fingerprints. Note that if some strains show identical fingerprints, indicating clonality, these will be considered as a single strain. If all strains of the same taxon show the same fingerprint, the description will be considered as an SSSD.   • All (non-clonal) strains described in the manuscript (or a subset if > 10 isolates) should be genome-sequenced in order to perform pairwise comparisons that will allow taxon circumscriptions.   • The monophyletic origin is a premise for recognizing a taxon. Therefore, all strains in the study must be placed within a phylogenetic framework. SAM requires that at least a 16S rRNA phylogenetic tree is provided, and the guidelines for this are given in the text.   • Genome coherence has to be evaluated using one of the OGRIs.   All strains must be phenotypically evaluated in order to reflect the intraspecific diversity.
Optional   The following are optional: • In cases of large collections (> 10 strains), and if well-justified, sequencing and phenotyping of only part of the collection will be allowed.   • Phylogenetic inference based on additional trees beyond the 16S rRNA gene, as indicated in the text, using different gene sets (or their translated products), such as ribosomal proteins, or core genes, is encouraged in order to further support and/or demonstrate the robustness of the classification.   • Chemotaxonomic markers should be determined for the whole collection, as appropriate.   • To increase the interest of the study, it is highly recommended to check the abundance of the new taxon in the sample of origin by means of culture-independent methods, such as fluorescence microscopy using specific probes (FISH), 16S rRNA gene amplicon sequencing, and/or by metagenomic read recruitments.   • To increase the importance of the study, SAM highly recommends screening for the presence and relative abundance of the new taxon in the publicly available 16S rRNA gene amplicon and metagenomic datasets. Read recruitments against the genome using the publicly available metagenomes can reveal relative abundance, biogeographic distribution and ecological patterns that increase the usefulness of the taxonomic description. For an example, see the study of the geographical distribution of *Hydrotalea lipotrueae*.

Table 1 contd. ...

...Table 1 contd.

Description of cultivated taxa and naming under the ICNP rules
Single strain species descriptions (SSSDs)
If only one isolate has been obtained, SAM will consider such manuscripts for review only under exceptional cases
Required • To compensate for the absence of multiple strains, SAM requires increasing the interest of the study by: • Providing a high-quality genome, ideally fully closed, and comparing the gene composition with the closest relatives. • Providing an exhaustive phenotypic description, ideally finding relevant traits in the genome and proving the expression of the capabilities. • Providing ecological information for the isolate by reporting its relative abundance in the sample of origin by means of culture-independent methods, such as FISH, 16S rRNA gene amplicon sequencing or metagenome read recruitments. • Providing the biogeographical distribution by means of screening for the presence and relative abundance of the new taxon in the publicly available 16S rRNA gene amplicon and metagenomic datasets. Read recruitments against the genome using the publicly available metagenomes can reveal relative abundance, biogeographic distribution and ecological patterns that increase the usefulness of the taxonomic description. For an example, see the study of the geographical distribution of *Hydrotalea lipotrueae* (Gao et al., 2021).
Optional Any additional information that increases the value of the contribution that goes beyond SSSDs will be welcomed, such as: • Evaluating the pan- and core-genomes of, for example, the genus where the strain has been classified. • Revealing taxonomically non-relevant traits, such as prophages, mobile elements and genomic islands that can increase the broader interest of the strain.

entirely from culture investigation, the utilization of isolates whole-genome provides appropriate species names that are derived entirely from culture evaluation. It is unethical to apply culture-based descriptions to individuals who are not cultured (Rosselló-Móra and Whitman, 2019). This is because it renders comments on the SeqCode in the part titled "threats" and "conclusions" of the SWOT analysis (Pallen, 2021) rendered ineffective. The information that is currently known as UnCode for prokaryotes that have not yet been cultured is referred to as the SeqCode, which is responsible for providing permanent names for the prokaryotes (Adopted from Willliam B. Whitman, Uncode Workshop, 2021). In order to differentiate it from the International Classification of New Prokaryotes (ICNP), which is used for the naming of all prokaryotes as long as the lower taxa (species and subspecies) are characterized by either a strain or a description. As stated in general consideration 7 of the SeqCode, it is composed of three primary components: the foundation of the code is referred to as Principles, the Rules are intended to put the Principles into action, and the Recommendations are intended to subsidise many of the Rules but do not have the same legal force as the Rules. In the year 2022,

Table 2. Requirements and recommendations for describing uncultivated taxa to be named under the SeqCode rules. The SeqCode is less restrictive than ICNP in linking a taxonomic description to a publication that is considered to be a permanent record. SeqCode requires that only new names are given in the publication proposing the name, and the full or succinct protologues, including the designated type material, can be given in the supplementary material. The SeqCode registry is available at (https://seqco.de/) and is the basis for name validation and priority establishment. SAM will follow these rules but endorse certain more restrictive requirements in order to ensure further (higher) data quality for the future. This table includes all details concerning data quality and reporting requirements, as well as recommendations for an isolate's genome, Metagenome Assembled Genome (MAG), or Single Amplified Genome (SAG) used to serve as the nomenclatural type for a species named under SeqCode, obtained largely from (Hedlund et al., 2022) (Permission for use of original content has been obtained by Dr. Bhagwan Rekadwad via CCC RightsLink; License Number: 5742870228563 dated Mar 06, 2024).

Description of taxa and naming under the SeqCode rules
Name and designated type genome
Required The taxon name and the designated type genome, with an INSDC accession number, need to be given in the main body of the manuscript or in a table if protologues are included in the supplementary material. Exceptionally, isolate-based descriptions with a nomenclature formulated under the SeqCode rules will be recognized by SAM if the strains, due to any specific reason, cannot be deposited in two different culture collections. Taxon names need to be written in italics and the genome designated names and accession numbers suffixed with a superscriptTS. For example, see Wolframiiraptor gerlachensis (Wger_A8TS = GCA_021323375.2TS).
Protologue Required The description of the taxon must be given either in the manuscript or in the supplementary material (if there are > 10 protologues) depending on the volume of descriptions given in the contribution. **The basic information required is:** • Name (formed with mnemonic cues). • Etymology. • Interpretation of the biological properties, inferred or demonstrated physiological traits, as well as ecological information, such as habitat, environmental physicochemical properties, inferred interactions with other organisms, such as symbiosis, syntrophy, and other properties that may be relevant for the description. • Designated genome assembly (e.g., INSDC accession) and access to raw data (e.g., SRA accession). • Given registry number after submission to the SeqCode Registry system. • Include as much metadata as possible (Field et al., 2008). Some fields of relevant metadata are already provided in the online template.
Optional Optional information can include: In cases where the authors wish to provide additional metadata fields, these can be added as new lines in the protologue template.

Table 2 contd. ...

...Table 2 contd.

Registry

Required

- Data quality will be assessed by using available automated pipelines or other approaches. Exceptions for lower data quality should be justified by the authors in the main body of the manuscript.
- The assembly should be available in one of the INSDC databases.
- Raw data (reads) should be available in the INSDC databases (e.g., Sequence Read Archive). This is not required for names effectively published before January 1, 2023, in order to allow for existing published names (e.g., existing *Candidatus* names) and names currently undergoing peer review to be validated under SeqCode.

Optional

There are several optional fields in the registry form which, if left empty/non-filled, will not lead to the invalidation of the entry. For details check (https://seqco.de/).

Type material data quality

Required

Genomes, MAGs and SAGs should have minimal standards of quality to be recognized as type material. Lower standards may lead to erroneous classifications and/or confusion that will promote undesired instability in the system. The major requirements for adequate (high) quality are:

- Assembly quality for MAGs and SAGs: > 90% complete and < 5% contaminated (modified from (Bowers et al., 2017)).
- For isolates as for MAGs/SAGs, read coverage ≥ 10x (Field et al., 2008).
- Provide evidence of the discreteness of the species, taxonomic rank, and position, including the uniqueness of the species with respect to existing validly named species, as well as justifying the taxonomic rank and position based on genomic and 16S rRNA gene data.
- Check for congruence between the genome-derived and 16S rRNA taxonomic assignments (e.g., (Karthikeyan et al., 2019)).
- For MAGs and SAGs, compare multiple high-quality genomes representing the species from more than one sample (multiple high-quality genomic assemblies from multiple samples can support the non-chimeric nature of MAGs and provide confidence of the assembly for both MAGs and SAGs) or justify why multiple samples and/or MAGs are not available.
- 16S rRNA gene > 75% complete, and passes chimera checks.
- > 80% of tRNAs present in order to further ensure high completeness of the genome (modified from (Bowers et al., 2017)).

Optional

Genomes, MAGs, and SAGs of high quality are desirable for representing a stable nomenclatural type. Therefore, we also suggest reporting:

- High genome integrity (contig # < 100; N50 > 25 kb; max. contig > 100 kb).

Please note that beyond the nomenclatural rules of the two codes, we also discuss below the standards required to classify taxa, even though the two represent distinct components of taxonomy. SAM is committed to the publication of excellence in taxonomy, and this discipline has three major components (Cowan, 1971): (i) Classification, which is the procedure for describing new taxa by exploring their diagnostic properties and placing them within the homonymous classification system that is expected to be operational, universal and predictive; (ii) Nomenclature, which is the only official discipline in taxonomy ruled by the ICNP and SeqCode codes, and pursues naming taxa with accurate rules and recommendations for a stable naming framework; and (iii) Identification, which is the "raison d'être" of taxonomy and supports accurately placing newly observed organisms as members of known taxa. Consequently, the editors will continue to uphold high standards of quality for genomes and taxonomic descriptions when editing submitted manuscripts.

Table 2 contd. ...

...Table 2 contd.

Most of the relevant requirements for taxonomic papers can be obtained in the Guide for Authors; in addition, and as guidelines for authors and reviewers, the editors want to emphasize that SAM:

✓ requires the almost complete 16S rRNA gene sequence (> 90% of total length) to be used for reconstructing the genealogical backbone of the organisms under study and recommends that this reconstruction represents the consensus topology for the evaluation of distinct trees generated using different algorithms (neighbour-joining, maximum parsimony and maximum likelihood), different sets of filters (depending on whether you are looking for ancient or modern lineages and filtering-out poorly aligned or hypervariable regions), and eventually different outgroups, and sets of sequences under study (i.e., trying to balance the number of reference and new sequences among the tree branches) in order to assess the robustness of the branching order. The LTP database (Ludwig et al., 2021), which contains all available type strain sequences, can be an excellent source of reference sequences for phylogenetic reconstructions with taxonomic purposes.

✓ encourages the use of genome sequences to select subsets of genes in order to reconstruct phylogenies. A reasonable practice will be to reconstruct trees based on core-genes or subsets of universal genes, such as the ribosomal proteins, single-copy universal genes, or the gene selection used by the Genome Taxonomy Database (GTDB; (Parks et al., 2018)).

✓ considers Multi Locus Sequence Analysis (MLSA) to be a robust tool for reconstructing protein-coding gene phylogenies but always recommends the use of a large number of genes (> 12; (Soria-Carrasco et al., 2007)) to ensure the stability of the branching orders.

✓ recommends evaluating the position of the new taxa using both concatenated alignment of protein-coding genes and 16S rRNA gene sequences, and reporting on the congruency (or lack) of the topologies of the resulting trees.

✓ requires genome sequencing of all strains in a study (or a subset of the strains if the collection is too large), at a high quality, not necessarily closed but with a low number of contigs and at least 10x coverage for each genome, in order to perform whole genome pairwise comparisons. Note that the SAM requirement of a sequenced genome for taxonomic purposes was implemented in 2016 and there have been no exceptions to this requirement in publications since then. Besides the Overall Genome Relatedness Index (OGRI; (Chun and Rainey, 2014)) parameters, SAM editors encourage the use of genomic information that goes beyond pure classification and helps in understanding intraspecific genetic diversity and biogeography, as well as inference of ecological and metabolic traits, among others.

✓ encourages the exploration of new methods to infer or determine the phenotype from genome sequences or other types of data. Since finding discriminative phenotypes may not always be straightforward, the use of commercial batteries of metabolic tests should be considered for isolate-based taxonomic descriptions. Chemotaxonomic characters, unless required by the respective taxonomic subcommittees or demonstrated to be highly discriminative for the taxon studied, may be regarded as optional, but the retrospective evaluation of such characters in a wider taxonomic framework is recommended (Sutcliffe et al., 2021).

✓ encourages the collection of a large number of strains, MAGs or SAGs, which are not clonal varieties of each other, since only this allows the genetic and phenotypic diversity to be assessed within a species.

Table 2 contd. ...

...Table 2 contd.

<table>
<tr><td>

✓ will only take Single Strain Species Descriptions (SSSDs) for publication under exceptional, well-justified cases, as shown in Table 1. These cases include, but are not limited to, fastidious microorganisms or exceptional metabolisms, taxonomic uniqueness or isolates that originate from extraordinary cultivation approaches. To increase the interest of the SSSDs contribution, we recommend linking them to metadata (e.g., in ecology using molecular approaches, such as metagenomics and single cell identification by fluorescence *in situ* hybridization). For example, the biogeographical distribution of a new species can be assessed by read recruitment plots using publicly available metagenomes (e.g., (Gago et al., 2021)).

✓ will only take MAGs and SAGs descriptions if these conform to the minimal requirements given in Table 2, or in well-justified cases when some of the minimum requirements are not met.

✓ requires the protologues to be published in the main body of the manuscript, as in Table 3.

</td></tr>
</table>

Table 3. Protologue table template (available online) to be used for a new genus and species. This information needs to be published as a table in the manuscript submitted to SAM, since the standard written descriptions are not necessary. This will be considered as the formal description for the name and, therefore, the effective publication of a new taxon (Permission for use of original content has been obtained by Dr. Bhagwan Rekadwad via CCC RightsLink; License Number: 5742870228563 dated Mar 06, 2024).

Note that SAM requires the inclusion of the accession number of the deposit of the genome sequence and 16S rRNA genes in one of the INSDC databases.

- All descriptions made following the nomenclatural rules of the ICNP must be given as a table in the main manuscript, as their validation in the IJSEM will require a permanent record and supplemental material is not considered (yet) to be permanent.

- The full descriptions made following the nomenclatural rules of the SeqCode can optionally be given in the supplemental material, since the permanent record is not a requirement for the SeqCode, although the registry system rules the priority and validation of the names. However, SAM requires that the name and designated type material are given in the main text. Therefore, in case of numerous taxa descriptions, SAM will consider publishing a supplementary table with all the complete protologues of the required information but, in addition, a single table with all new names and their designated type material must be given within the main body of the manuscript. For example, see *Wolframiiraptor gerlachensis* (Wger_A8[TS] = GCA_021323375.2[TS]).

- In the protologue table, [req] indicates required and [opt] indicates recommended but optional. [req under ICNP] indicates the requirements that have to be taken into account when naming under the ICNP rules, and [req under SeqCode] indicates the requirements that have to be taken into account when naming under the SeqCode rules.

Guiding Code for Nomenclature [req]	Indicate whether the rules of the **ICNP** or **SeqCode** have been followed.
Nature of the type material [req]	Indicate the **strain** or **genome sequence**
Genus name [req]	Add the generic name (e.g., *Salinibacter*)

Table 3 contd. ...

...Table 3 contd.

Species name [req]	Add the full name (e.g., *Salinibacter altiplanensis*)
Genus status [req]	Select the adequate status **(gen. nov./nom. rev.)**
Genus etymology [req]	**[N.B. Genus etymology is only for new genus proposals; never for extant genera]** Add the genus etymology (e.g., Sa.li.ni.bac'ter. L. fem. pl. n. *salinae* salterns, salt-works; N.L. masc. n. *bacter* masc. equivalent of the Gr. neut. n. *bakterion* a rod; N.L. masc. n. *Salinibacter* a rod from salt-works).
Type species of the genus [req]	**[N.B. Type species of the genus is only for new genus proposals; never for extant genera]** Indicate the designated type species of the genus by adding the full specific name (e.g., *Salinibacter ruber*).
Specific epithet [req]	Add the specific epithet (e.g., *altiplanensis*).
Species status [req]	Select the adequate status **(sp. nov./comb. nov./nom. rev.)**
Species etymology [req]	Add the specific epithet etymology (e.g., al.ti.pla.ne'nsis, N.L. masc. adj. *altiplanensis* of the Argentinian Altiplano).
Designation of the Type Strain [req under ICNP]	Provide the designation of the strain as in the manuscript (e.g., AN15^T). Note that a superscript T must follow the strain designation(s).
Strain Collection Numbers [req under ICNP]	Provide at least two collection numbers of two international repositories (e.g., IBRC-M 11031^T = CECT 9105^T). Note that a superscript T must follow the strain designation(s).
Designated Genome, MAG, or SAG [req under SeqCode]	Provide the designation of the genome, MAG or SAG as in the manuscript (e.g., Wger_A8^{TS}). Note that a superscript TS must follow the genome designation(s)
Type Genome, MAG, or SAG accession Nr. [INSDC databases] [req under SeqCode]	Add the genome accession number(s) with the repository identifier (e.g., GenBank = GCA_021323375.2^{TS}). Note that a superscript TS must follow the genome designation(s) if following the SeqCode rules. If following the ICNP rules, no suffix is needed.
Access to raw data (e.g., SRA accession) [opt under SeqCode]	Add the accession number to the raw data from where the MAGs were assembled and binned (required for names after January 1, 2023).
Registry number [req under SeqCode]	Add the registry number available at (https://disc-genomics. uibk.ac.at/seqcode/).
Genome status [opt]	Complete/incomplete.
Genome size [opt]	Add the genome size in kbp (e.g., 3,580).
GC mol% [opt]	Add the GC mol% (e.g., 64.41).
16S rRNA gene accession nr. [req]	Provide the gene accession number(s) (e.g., LT160741).

Table 3 contd. ...

...Table 3 contd.

Description of the new taxon and diagnostic traits [req]	Provide the phenotypic, genotypic and, if possible, ecological description of the new taxon emphasizing the traits used to diagnose it. This text can be as long as the normal written protologues in standard publications. In addition, add data here on chemotaxonomy, if determined.
Country of origin [opt]	Add the country of origin (e.g., Argentina).
Region of origin [opt]	Add the region of origin of isolation (e.g., Salar de Antofalla).
Date of isolation (dd/mm/yyyy) [opt]	Add the isolation date.
Source of isolation [opt]	Add the source of isolation (e.g., hypersaline lake).
Sampling date (dd/ mm/yyyy) [opt]	Add the sampling date.
Latitude (xx°xx'xx"N/S) [opt]	Add the latitude coordinates.
Longitude (xx°xx'xx"E/W) [opt]	Add the longitude coordinates.
Altitude (metres above sea level) [opt]	Add the altitude in metres.
Number of strains in study [opt]	Add the number of strains in the study.
Source of isolation of non-type strains [opt]	Add the source of isolation of the additional strains.
Information related to the Nagoya Protocol [req]	Add here the permit number(s) and country of origin and/or local authority from where the permits had been obtained.

- If you are publishing a new species in an existing genus with a validly published name, the second column must be the description of this new taxon, and the fields *Genus status*, *Genus etymology* and *Type species of the genus* can be removed. If you have more than one species, you only need to add a third, fourth, fifth ... column - one for each description.

- If you are publishing a new genus in addition to a new species, the second column must be the description of the genus, and then the fields ranging from *Specific epithet* to *GC mol%* may be filled in with a dash indicating an empty field. The genus description must be followed by the type species of the genus in a third column, and additional species will follow in the fourth, fifth ... columns. Should you have two or more genera, just provide the Genus, Type species, Species, ... repeating the column series or, alternatively, you can generate one table for each different genus.

Table 3 contd. ...

...Table 3 contd.

- Please add a superscript T to the designated type strain name and strain collection numbers (e.g., AN5[T]; IBRC-M 11031[T] = CECT 9105[T]) if you name under the ICNP rules.
- Please add a superscript TS to the designated type genome name and strain collection numbers (e.g., Wger_A8[TS] = GCA_021323375.2[TS]) if you name under the SeqCode rules.
- Empty rows (i.e., fields where no information is available) must be removed.

the SeqCode was put into effect. Additionally, additional information regarding the SeqCode can be found on the following website: SeqCode Registry (https://disc-genomics.uibk.ac.at/seqcode/). Details of rules SeqCode are also given in Tables 1–3.

Acknowledgements

The author duly acknowledges content as per CC BY 4.0 from (Rekadwad, 2024). Permission for use of original content has been obtained by Dr. Bhagwan Rekadwad via CCC RightsLink; License Number: 5742870228563 dated Mar 06, 2024.

References

Alneberg, J., Karlsson, C. M. G., Divne, A. M., Bergin, C., Homa, F., Lindh, M. V. et al. (2018). Genomes from uncultivated prokaryotes: a comparison of metagenome-assembled and single-amplified genomes. Microbiome 6(1): 173. https://doi.org/10.1186/s40168-018-0550-0.

Bowers, R. M., Kyrpides, N. C., Stepanauskas, R., Harmon-Smith, M., Doud, D., Reddy, T. B. K. et al. (2017). Minimum information about a single amplified genome (MISAG) and a metagenome-assembled genome (MIMAG) of bacteria and archaea. Nature Biotechnology 35(8): 725–731. https://doi.org/10.1038/nbt.3893.

Chun, J. and Rainey, F. A. (2014). Integrating genomics into the taxonomy and systematics of the Bacteria and Archaea. International Journal of Systematic and Evolutionary Microbiology 64(Pt 2): 316–324. https://doi.org/10.1099/ijs.0.054171-0.

Cowan, S. T. (1965). Principles and Practice of bacterial taxonomy—a forward look. J. Gen. Microbiol. 39: 148–158.

Cowan, S. T. (1971). Sense and nonsense in bacterial taxonomy. Journal of General Microbiology 67(1): 1–8. https://doi.org/10.1099/00221287-67-1-1.

Field, D., Garrity, G., Gray, T., Morrison, N., Selengut, J., Sterk, P. et al. (2008). The minimum information about a genome sequence (MIGS) specification. Nature Biotechnology 26(5): 541–547. https://doi.org/10.1038/nbt1360.

Gago, J. F., Viver, T., Urdiain, M., Pastor, S., Kämpfer, P., Robledo, P. A. et al. (2021). Comparative genome analysis of the genus Hydrotalea and proposal of the novel species *Hydrotalea lipotrueae* sp. nov., isolated from a groundwater aquifer in the south of Mallorca Island, Spain. Systematic and Applied Microbiology 44(6): 126277. https://doi.org/10.1016/j.syapm.2021.126277.

Hedlund, B. P., Chuvochina, M., Hugenholtz, P., Konstantinidis, K. T., Murray, A. E., Palmer, M. et al. (2022). SeqCode: a nomenclatural code for prokaryotes described from sequence data. Nature Microbiology 7(10): 1702–1708. https://doi.org/10.1038/s41564-022-01214-9.

Hug, L. A., Baker, B. J., Anantharaman, K., Brown, C. T., Probst, A. J., Castelle, C. J. et al. (2016). A new view of the tree of life. Nature Microbiology 1: 16048. https://doi.org/10.1038/nmicrobiol.2016.48.

Karthikeyan, S., Rodriguez-R, L. M., Heritier-Robbins, P., Kim, M., Overholt, W. A., Gaby, J. C. et al. (2019). "Candidatus Macondimonas diazotrophica", a novel gammaproteobacterial genus dominating crude-oil-contaminated coastal sediments. The ISME Journal 13(8): 2129–2134. https://doi.org/10.1038/s41396-019-0400-5.

Konstantinidis, K. T., Rosselló-Móra, R. and Amann, R. (2017). Uncultivated microbes in need of their own taxonomy. The ISME Journal 11(11): 2399–2406. https://doi.org/10.1038/ismej.2017.113.

Ludwig, W., Viver, T., Westram, R., Francisco Gago, J., Bustos-Caparros, E., Knittel, K. et al. (2021). Release LTP_12_2020, featuring a new ARB alignment and improved 16S rRNA tree for prokaryotic type strains. Systematic and Applied Microbiology 44(4): 126218. https://doi.org/10.1016/j.syapm.2021.126218.

Mukherjee, S., Seshadri, R., Varghese, N. J., Eloe-Fadrosh, E. A., Meier-Kolthoff, J. P., Göker, M. et al. (2017). 1,003 reference genomes of bacterial and archaeal isolates expand coverage of the tree of life. Nature Biotechnology 35(7): 676–683. https://doi.org/10.1038/nbt.3886.

Murray, R. G. and Schleifer, K. H. (1994). Taxonomic notes: a proposal for recording the properties of putative taxa of procaryotes. International Journal of Systematic Bacteriology 44(1): 174–176. https://doi.org/10.1099/00207713-44-1-174.

Oren, A. and Garrity, G. M. (2018). List of new names and new combinations previously effectively, but not validly, published. International Journal of Systematic and Evolutionary Microbiology 68(11): 3379–3393. https://doi.org/10.1099/ijsem.0.003071.

Pallen, M. J. (2021). The status Candidatus for uncultured taxa of Bacteria and Archaea: SWOT analysis. International Journal of Systematic and Evolutionary Microbiology 71(9): 005000. https://doi.org/10.1099/ijsem.0.005000.

Parks, D. H., Rinke, C., Chuvochina, M., Chaumeil, P. A., Woodcroft, B. J., Evans, P. N. et al. (2017). Recovery of nearly 8,000 metagenome-assembled genomes substantially expands the tree of life. Nature Microbiology 2(11): 1533–1542. https://doi.org/10.1038/s41564-017-0012-7.

Parks, D. H., Chuvochina, M., Waite, D. W., Rinke, C., Skarshewski, A., Chaumeil, P. A. et al. (2018). A standardized bacterial taxonomy based on genome phylogeny substantially revises the tree of life. Nature Biotechnology 36(10): 996–1004. https://doi.org/10.1038/nbt.4229.

Parks, D. H., Chuvochina, M., Chaumeil, P. A., Rinke, C., Mussig, A. J. and Hugenholtz, P. (2020). A complete domain-to-species taxonomy for Bacteria and Archaea. Nature Biotechnology 38(9): 1079–1086. https://doi.org/10.1038/s41587-020-0501-8.

Rekadwad, B. N. (2024). Taxonomic approaches for uncultivated prokaryotes. Li, W. J., Jiao, J. Y., Salam, N. and Narsing Rao, M. P. (eds.). Modern Taxonomy of Bacteria and Archaea: New Methods, Technology and Advances. Springer Nature. ISBN: 978-981-99-5719-4. https://doi.org/10.1007/978-981-99-5720-0_8.

Rodriguez-R, L. M., Gunturu, S., Harvey, W. T., Rosselló-Mora, R., Tiedje, J. M., Cole, J. R. et al. (2018a). The Microbial Genomes Atlas (MiGA) webserver: taxonomic and gene diversity analysis of Archaea and Bacteria at the whole genome level. Nucleic Acids Research 46(W1): W282–W288. https://doi.org/10.1093/nar/gky467.

Rodriguez-R, L. M., Castro, J. C., Kyrpides, N. C., Cole, J. R., Tiedje, J. M. and Konstantinidis, K. T. (2018b). How Much Do rRNA gene surveys underestimate extant bacterial diversity?. Applied and Environmental Microbiology 84(6): e00014–18. https://doi.org/10.1128/AEM.00014-18.

Rosselló-Móra, R. and Whitman, W. B. (2019). Dialogue on the nomenclature and classification of prokaryotes. Systematic and Applied Microbiology 42(1): 5–14. https://doi.org/10.1016/j.syapm.2018.07.002.

Sikorski, J., Rosselló-Mora, R. and Lorenz, M. G. (1999). Analysis of genotypic diversity and relationships among Pseudomonas stutzeri strains by PCR-based genomic fingerprinting and multilocus enzyme electrophoresis. Systematic and Applied Microbiology 22(3): 393–402. https://doi.org/10.1016/S0723-2020(99)80048-4.

Soria-Carrasco, V., Valens-Vadell, M., Peña, A., Antón, J., Amann, R., Castresana, J. et al. (2007). Phylogenetic position of Salinibacter ruber based on concatenated protein alignments. Systematic and Applied Microbiology 30(3): 171–179. https://doi.org/10.1016/j.syapm.2006.07.001.

Sutcliffe, I. C., Rosselló-Móra, R. and Trujillo, M. E. (2021). Addressing the sublime scale of the microbial world: reconciling an appreciation of microbial diversity with the need to describe species. New Microbes and New Infections 43: 100931. https://doi.org/10.1016/j.nmni.2021.100931.

Chapter 8

Quantitative Insights into Microbial Ecology Version 2.0 (QIIME2) for End-to-End Analysis of Metagenome Sequence Data by /Bolyen et al., 2019 (#As it is)

Introduction

Understanding of the microbial world has significantly advanced in the last 20 years thanks to the rapid advancements in DNA sequencing and bioinformatics technologies. The vast diversity of microorganisms, how microbiota and microbiomes affect disease (Smith et al., 2013), how medical treatment (Gopalakrishnan et al., 2018) and how microorganisms affect the planet's health (Gehring et al., 2017), and the emerging field of microbiome biotechnology's applications in medicine (Lee et al., 2018), forensics (Metcalf et al., 2016), the environment (Pineda et al., 2017), and agriculture (Lee et al., 2018) are all subjects of this growing understanding. A significant portion of this research has been fuelled by marker-gene surveys, which characterize the microbiota with differing levels of taxonomic precision and phylogenetic information. Examples of these surveys include fungal internal-transcribed-spacer sections, bacterial/archaeal 16S rRNA genes, and eukaryotic 18S rRNA genes. Other data types, such metabolite (Kapono et al., 2018), metaproteome

(Verberkmoes et al., 2009), or metatranscriptome (Verberkmoes et al., 2009; Barr et al., 2018) profiles, are currently being integrated into the field.

Many microbiome investigations have been enabled by the QIIME 1 microbiome bioinformatics platform, which has also attracted a sizable user and development community. Our workshops, direct collaborations, and interactions with QIIME 1 users in our online support forum have demonstrated the platform's capacity to support an ever-widening range of microbiome researchers in academia, government, and industry. Here, we introduce QIIME 2, a fully redesigned and rebuilt system that will hopefully enable the next generation of microbiome study by enabling reproducible and modular analysis of microbiome data.

Quantitative Insights Into Microbial Ecology 2 (QIIME 2) is a program that was developed for the purpose of evaluating and interpreting sequencing data pertaining to microorganisms and succeeded QIIME 1. Data preparation, diversity analysis, taxonomic assignment, and statistical analysis are just some of the activities that may be accomplished with the help of the various tools and algorithms that are available in QIIME 2 (Table 1). The scientific community makes extensive use of it for the purpose of researching microbial communities in a variety of settings, such as soil, water, and the human body. The second version of QIIME is a free and open-source software that was built by a community of researchers and developers working together (Bolyen et al., 2019). QIIME 2 can be used to analyze any kind of microbiome data, for instance, community characterization of soil from the Central Deccan Plateau dry tropical deciduous forest (Ceddtrof) in central India (Rekadwad et al., 2024).

Table 1. Difference between QIIME1 and QIIME2.

Feature	QIIME 1	QIIME 2
Framework	Collection of Python scripts	Plugin-based architecture
OTU Clustering	Open-reference, closed-reference, or *de novo*	Denoising (ASVs)
Installation & Use	Complex, requires expertise	Streamlined, user-friendly interface

"Moving Pictures" tutorial

This tutorial uses QIIME 2 to perform an analysis of human microbiome samples from two individuals at four body sites at five timepoints, the first of which immediately followed antibiotic usage. A study based on these samples was originally published in Caporaso et al. (2010). The data used in this tutorial were sequenced on an Illumina HiSeq using the Earth Microbiome Project hypervariable region 4 (V4) 16S rRNA sequencing protocol (Caporaso et al., 2010).

Before beginning this tutorial, create a new directory and change to that directory.

mkdir qiime2-moving-pictures-tutorial
cd qiime2-moving-pictures-tutorial

Sample metadata

Before starting the analysis, explore the sample metadata so that one can familiarize themselves with the samples used for this analysis. The sample metadata is available as a Google Sheet. One can download this file as tab-separated text by selecting File > Download as > Tab-separated values. Alternatively, the following command will download the sample metadata as tab-separated text and save it in the file sample-metadata.tsv. This sample-metadata.tsv file is used throughout the rest of the tutorial.

Please select a download option that is most appropriate for your environment:

*wget *
  *-O "sample-metadata.tsv" *
  "https://data.qiime2.org/2024.2/tutorials/moving-pictures/sample_metadata.
  tsv"

Obtaining and importing data

Make a directory for sequence and download the sequence QIIME 2 artifacts single-end or paired-end using *wget* command as per requirement.

mkdir emp-single-end-sequences

*wget *
  *-O "emp-single-end-sequences/barcodes.fastq.gz" *
  "https://data.qiime2.org/2024.2/tutorials/moving-pictures/emp-single-end-
  sequences/barcodes.fastq.gz"

OR

*wget *
  *-O "emp-single-end-sequences/sequences.fastq.gz" *
  "https://data.qiime2.org/2024.2/tutorials/moving-pictures/emp-single-end-
  sequences/sequences.fastq.gz"

The next step is to import downloaded sequences to bash.

*qiime tools import *
*--type EMPSingleEndSequences *
*--input-path emp-single-end-sequences *
--output-path emp-single-end-sequences.qza

Output artifacts:

emp-single-end-sequences.qza

It is possible to check the UUID, type, and format of your newly-imported sequences, confirming that your import worked as expected:

qiime tools peek emp-single-end-sequences.qza

stdout:

UUID	*: 2d89a678-ef53-4f36-bab3-10d63ee4b28c*
Type	*: EMPSingleEndSequences*
Data format	*: EMPSingleEndDirFmt*

Demultiplexing sequences

To demultiplex sequences we need to know which barcode sequence is associated with each sample. This information is contained in the sample metadata file. Run the following commands to demultiplex the sequences. The demux.qza QIIME 2 artifact will contain the demultiplexed sequences. The second output (demux-details.qza) presents Golay error correction details, and will not be explored in this tutorial.

*qiime demux emp-single *
   *--i-seqs emp-single-end-sequences.qza *
   *--m-barcodes-file sample-metadata.tsv *
   *--m-barcodes-column barcode-sequence *
   *--o-per-sample-sequences demux.qza *
   --o-error-correction-details demux-details.qza

Output artifacts:

demux-details.qza
demux.qza

After demultiplexing, it is useful to generate a summary of the demultiplexing results. This allows to determine how many sequences were obtained per sample, and also to get a summary of the distribution of sequence qualities at each position in your sequence data.

*qiime demux summarize *
   *--i-data demux.qza *
   --o-visualization demux.qzv

Output visualizations:

demux.qzv

Note: All QIIME 2 visualizers (i.e., commands that take a --o-visualization parameter) will generate a .qzv file. These files can be viewed with *qiime tools view* command.,

qiime tools view demux.qzv

Alternatively, QIIME 2 artifacts and visualizations at view.qiime2.org by uploading files or providing URLs.

Sequence quality control and feature table construction

Several quality control techniques, such as DADA2, Deblur, and simple quality-score-based filtering, are available as plugins for QIIME 2. We now introduce Deblur and DADA2. A FeatureTable[Frequency] QIIME 2 artifact, which contains counts (frequencies) of each unique sequence in each sample in the dataset, and a FeatureData[Sequence] QIIME 2 artifact, which maps feature identifiers in the FeatureTable to the sequences they represent, are produced by both DADA2 and Deblur, from which the user can select any option for their analysis.

Option 1: DADA2

Certain samples might have excellent starting base quality, removing any bases from the start of the sequences. About base 120, the quality starts to deteriorate; sequences at that base are truncated. The slowest command in this lesson could take up to 10 minutes to execute.

*qiime dada2 denoise-single *
  *--i-demultiplexed-seqs demux.qza *
  *--p-trim-left 0 *
  *--p-trunc-len 120 *
  *--o-representative-sequences rep-seqs-dada2.qza *
  *--o-table table-dada2.qza *
  --o-denoising-stats stats-dada2.qza

Output artifacts:

stats-dada2.qza
table-dada2.qza
rep-seqs-dada2.qza

*qiime metadata tabulate *
  *--m-input-file stats-dada2.qza *
  --o-visualization stats-dada2.qzv

Output visualizations:

stats-dada2.qzv

If you would like to continue the tutorial using this FeatureTable (opposed to the Deblur feature table generated in Option 2), run the following commands.

mv rep-seqs-dada2.qza rep-seqs.qza
mv table-dada2.qza table.qza

Output artifacts:

rep-seqs.qza
table.qza

Option 2: Deblur

Deblur uses sequence error profiles to associate erroneous sequence reads with the true biological sequence from which they are derived, resulting in high quality sequence variant data. This is applied in two steps. First, an initial quality filtering process based on quality scores is applied. This method is an implementation of the quality filtering approach described by Bokulich et al. (2013).

*qiime quality-filter q-score *
*   --i-demux demux.qza *
*   --o-filtered-sequences demux-filtered.qza *
*   --o-filter-stats demux-filter-stats.qza*

Output artifacts:

demux-filtered.qza
demux-filter-stats.qza

Next, the Deblur workflow is applied using the qiime deblur denoise-16S method. This method requires one parameter that is used in quality filtering, --p-trim-length n which truncates the sequences at position n. In general, the Deblur developers recommend setting this value to a length where the median quality score begins to drop too low. On these data, the quality plots (prior to quality filtering) suggest a reasonable choice is in the 115 to 130 sequence position range. This is a subjective assessment. One situation where you might deviate from that recommendation is when performing a meta-analysis across multiple sequencing runs. In this type of meta-analysis, it is critical that the read lengths be the same for all of the sequencing runs being compared to avoid introducing a study-specific bias. Since we already using a trim length of 120 for qiime dada2 denoise-single, and since 120 is reasonable given the quality plots, we will pass --p-trim-length 120. This next command may take up to 10 minutes to run.

*qiime deblur denoise-16S *
  *--i-demultiplexed-seqs demux-filtered.qza *
  *--p-trim-length 120 *
  *--o-representative-sequences rep-seqs-deblur.qza *
  *--o-table table-deblur.qza *
  *--p-sample-stats *
  --o-stats deblur-stats.qza

Output artifacts:

deblur-stats.qza
table-deblur.qza

rep-seqs-deblur.qza
*qiime metadata tabulate *
  *--m-input-file demux-filter-stats.qza *
  --o-visualization demux-filter-stats.qzv
*qiime deblur visualize-stats *
  *--i-deblur-stats deblur-stats.qza *
  --o-visualization deblur-stats.qzv

Output visualizations:

demux-filter-stats.qzv
deblur-stats.qzv

If you would like to continue the tutorial using this FeatureTable (opposed to the DADA2 feature table generated in Option 1), run the following commands.

mv rep-seqs-deblur.qza rep-seqs.qza
mv table-deblur.qza table.qza

Feature Table and Feature Data summaries

Visual summaries of the resulting data allow for exploration of the data after the quality filtering stage is finished. Histograms of those distributions, some associated summary statistics, and the number of sequences corresponding to each sample and feature will all be provided by the feature-table summarize command. A mapping of feature IDs to sequences and links to quickly BLAST each sequence against the NCBI nt database are provided by the feature-table tabulate-seqs tool. Later in the tutorial, the latter visualization will come in very handy to gain further insight into particular aspects that are significant in the data set.

```
qiime feature-table summarize \
   --i-table table.qza \
   --o-visualization table.qzv \
   --m-sample-metadata-file sample-metadata.tsv
qiime feature-table tabulate-seqs \
   --i-data rep-seqs.qza \
   --o-visualization rep-seqs.qzv
```

Output visualizations:

table.qzv
rep-seqs.qzv

Generate a tree for phylogenetic diversity analyses

Several phylogenetic diversity measures are supported by QIIME, such as weighted and unweighted UniFrac and Faith's Phylogenetic Diversity. A rooted phylogenetic tree linking the features to each other is necessary for these metrics, in addition to counts of features per sample (i.e., the data in the FeatureTable [Frequency] QIIME 2 artifact). A Phylogeny [Rooted] QIIME 2 artifact will have this data. With the q2-phylogeny plugin, we will use the align-to-tree-mafft-fasttree process to create a phylogenetic tree.

Feature Data [Aligned Sequence] QIIME 2 artifact is produced by the pipeline first using the mafft programme to conduct a multiple sequence alignment of the sequences in our Feature Data [Sequence]. The pipeline then masks (or filters) the alignment to eliminate highly changeable places. These places are thought to introduce noise into the final phylogenetic tree. Next, using the masked alignment, the pipeline uses Fast Tree to create a phylogenetic tree. The Fast Tree programme produces an unrooted tree; hence, the last step in this section is applying midpoint rooting to position the tree's root at the middle of the unrooted tree's longest tip-to-tip distance.

```
qiime phylogeny align-to-tree-mafft-fasttree \
   --i-sequences rep-seqs.qza \
   --o-alignment aligned-rep-seqs.qza \
   --o-masked-alignment masked-aligned-rep-seqs.qza \
   --o-tree unrooted-tree.qza \
   --o-rooted-tree rooted-tree.qza
```

Output artifacts:

aligned-rep-seqs.qza
masked-aligned-rep-seqs.qza
rooted-tree.qza
unrooted-tree.qza

Alpha and beta diversity analysis

The q2-diversity plugin, which facilitates the computation of alpha and beta diversity measures, the application of pertinent statistical tests, and the creation of interactive visualizations, provides access to QIIME 2's diversity studies. The core-metrics-phylogenetic approach, which uses Emperor to create principal coordinates analysis (PCoA) plots for each of the beta diversity metrics, computes multiple alpha and beta diversity metrics, and rarefies a Feature Table [Frequency] to a depth that the user specifies. The default metrics that are computed are:

Alpha diversity

- Shannon's diversity index (a quantitative measure of community richness)
- Observed Features (a qualitative measure of community richness)
- Faith's Phylogenetic Diversity (a qualitative measure of community richness that incorporates phylogenetic relationships between the features)
- Evenness (or Pielou's Evenness; a measure of community evenness)

Beta diversity

- Jaccard distance (a qualitative measure of community dissimilarity)
- Bray-Curtis distance (a quantitative measure of community dissimilarity)
- Unweighted UniFrac distance (a qualitative measure of community dissimilarity that incorporates phylogenetic relationships between the features)
- Weighted UniFrac distance (a quantitative measure of community dissimilarity that incorporates phylogenetic relationships between the features)

The even sampling (or rarefaction) depth, or --p-sampling-depth, is a crucial parameter that must be entered into this script. This script will randomly subsample the counts from each sample to the parameter value because most diversity metrics are sensitive to varying sampling depths across various samples. To ensure that every sample in the final table has a total count of 500, this step will subsample the counts in each sample without replacement, for instance, if you supply --p-sampling-depth 500. Samples that have a total count below this threshold will not be included in the diversity analysis. Selecting this number is challenging. We advise examining the data in the table before making a decision to the above-created table.qzv file. Select the highest number that will allow you to keep more sequences per sample while removing the fewest samples possible.

*qiime diversity core-metrics-phylogenetic *
  *--i-phylogeny rooted-tree.qza *
  *--i-table table.qza *
  *--p-sampling-depth 1103 *
  *--m-metadata-file sample-metadata.tsv *
  --output-dir core-metrics-results

Output artifacts:

core-metrics-results/faith_pd_vector.qza
core-metrics-results/unweighted_unifrac_distance_matrix.qza
core-metrics-results/bray_curtis_pcoa_results.qza
core-metrics-results/shannon_vector.qza
core-metrics-results/rarefied_table.qza
core-metrics-results/weighted_unifrac_distance_matrix.qza
core-metrics-results/jaccard_pcoa_results.qza
core-metrics-results/weighted_unifrac_pcoa_results.qza
core-metrics-results/observed_features_vector.qza
core-metrics-results/jaccard_distance_matrix.qza
core-metrics-results/evenness_vector.qza
core-metrics-results/bray_curtis_distance_matrix.qza
core-metrics-results/unweighted_unifrac_pcoa_results.qza

Output visualizations:

core-metrics-results/unweighted_unifrac_emperor.qzv
core-metrics-results/jaccard_emperor.qzv
core-metrics-results/bray_curtis_emperor.qzv
core-metrics-results/weighted_unifrac_emperor.qzv

The --p-sampling-depth parameter is now set to 1103. Based on the number of sequences in the L3S313 sample, this value was selected because it is both significantly higher (relatively) than the number of sequences in the samples with fewer sequences and close to the number of sequences in the next few samples that have higher sequence counts. As a result, we will be able to keep most of our samples. The three samples with the lowest sequence counts will not be included in any core-metrics, phylogenetic, or other analysis that makes use of these findings. It is important to note that these three samples are all "right palm" samples. It is not ideal to lose an excessive amount of samples from one metadata category. This felt like the best compromise between the total sequences analyzed and the number of samples retained, though, as we are dropping a sufficient number of samples.

Investigate the microbial makeup of the samples in relation to the sample metadata after computing diversity metrics. The sample metadata file that was previously downloaded contains this data. First, we will look for correlations between the alpha diversity data and the category metadata columns. Here, we will do that for the evenness and Faith Phylogenetic Diversity metrics, which gauge community richness.

*qiime diversity alpha-group-significance \
--i-alpha-diversity core-metrics-results/faith_pd_vector.qza \
--m-metadata-file sample-metadata.tsv \
--o-visualization core-metrics-results/faith-pd-group-significance.qzv*

*qiime diversity alpha-group-significance \
--i-alpha-diversity core-metrics-results/evenness_vector.qza \
--m-metadata-file sample-metadata.tsv \
--o-visualization core-metrics-results/evenness-group-significance.qzv*

Output visualizations:

core-metrics-results/faith-pd-group-significance.qzv
core-metrics-results/evenness-group-significance.qzv

There is no correlation between alpha diversity and any of the continuous sample metadata columns in this data set (like days-since-experiment-start), so we will not be testing for those associations here. It can be carried out with the qiime diversity alpha-correlation command (for this data set or other data sets). The sample composition will be analyzed using the beta-group-significance command in the context of categorical metadata using PERMANOVA, as described by Anderson (2017). The ensuing instructions will ascertain whether sample distances within a group, samples from the same body site are more similar to one another than they are to samples from other groups, tongue, left palm, and right palm. In the event that the beta-group-significance command is invoked with the --p-pairwise parameter, pairwise tests will be conducted to ascertain whether any particular pairs of groups (such as tongue and gut) differ from one another. As this command is based on permutation tests, it may take a while to execute, particularly when passing --pairwise. As such, running beta-group-significance on particular metadata columns will be investigated instead of all metadata columns to which it is applicable, in contrast to the previous commands. This will apply in this case to unweighted UniFrac distances using the following two sample metadata columns.

qiime diversity beta-group-significance \
  --i-distance-matrix   core-metrics-results/unweighted_unifrac_distance_matrix.
  qza \
  --m-metadata-file sample-metadata.tsv \
  --m-metadata-column body-site \
  --o-visualization   core-metrics-results/unweighted-unifrac-body-site-
  significance.qzv \
  --p-pairwise

qiime diversity beta-group-significance \
  --i-distance-matrix   core-metrics-results/unweighted_unifrac_distance_matrix.
  qza \
  --m-metadata-file sample-metadata.tsv \
  --m-metadata-column subject \
  --o-visualization   core-metrics-results/unweighted-unifrac-subject-group-
  significance.qzv \
  --p-pairwise

Output visualizations:

core-metrics-results/unweighted-unifrac-body-site-significance.qzv
core-metrics-results/unweighted-unifrac-subject-group-significance.qzv

Optionally, for determining the correlation between sample metadata and sample composition, use the qiime metadata distance matrix in combination with the qiime diversity mantel and qiime diversity bioenv commands.

Finally, ordination is a popular approach for exploring microbial community composition in the context of sample metadata. We can use the Emperor tool to explore principal coordinates (PCoA) plots in the context of sample metadata. While our core-metrics-phylogenetic command did already generate some Emperor plots, we want to pass an optional parameter, --p-custom-axes, which is very useful for exploring time series data. The PCoA results that were used in core-metrics-phylogeny are also available, making it easy to generate new visualizations with Emperor. We will generate Emperor plots for unweighted UniFrac and Bray-Curtis so that the resulting plot will contain axes for principal coordinate 1, principal coordinate 2, and days since the experiment start. Use that last axis to explore how these samples changed over time.

At last, ordination is a well-liked method for examining the makeup of the microbial community in relation to sample metadata. Principal coordinates (PCoA) plots can be examined using the Emperor tool in relation to sample metadata. Although some Emperor plots were already produced by our *core-metrics-phylogenetic* command, if you still want to pass the optional parameter *--p-custom-axes*, which is very helpful for examining time series data. With the availability of the PCoA results used in *core-metrics-phylogeny*, creating new visualizations with Emperor is a breeze. For unweighted UniFrac and Bray-Curtis, create Emperor plots will have axes for principal coordinate 1, principal coordinate 2, and the number of days since the experiment began. Examine how these samples changed over time using that final axis.

*qiime emperor plot *
    *--i-pcoa core-metrics-results/unweighted_unifrac_pcoa_results.qza *
    *--m-metadata-file sample-metadata.tsv *
    *--p-custom-axes days-since-experiment-start *
    --o-visualization   core-metrics-results/unweighted-unifrac-emperor-days-since-experiment-start.qzv

*qiime emperor plot *
    *--i-pcoa core-metrics-results/bray_curtis_pcoa_results.qza *
    *--m-metadata-file sample-metadata.tsv *
    *--p-custom-axes days-since-experiment-start *
    --o-visualization   core-metrics-results/bray-curtis-emperor-days-since-experiment-start.qzv

Output visualizations:

core-metrics-results/bray-curtis-emperor-days-since-experiment-start.qzv
core-metrics-results/unweighted-unifrac-emperor-days-since-experiment-start.qzv

Alpha rarefaction plotting

The *qiime diversity alpha-rarefaction* visualizer can be used to investigate alpha diversity as a function of sampling depth. The alpha diversity metrics are calculated by this visualizer at various sampling depths, in steps ranging from 1 (which can be adjusted using *--p-min-depth*) to the value entered as *--p-max-depth*. Ten rarefied tables will be produced at each sampling depth step, and for each sample in each table, the diversity metrics will be calculated. With *--p-iterations*, the number of iterations (rarefied tables computed at each sampling depth) can be adjusted. If the sample metadata is supplied with the *--m-metadata-file* option, samples can be grouped according to the metadata in the resulting visualization. Average diversity values will be plotted for each sample at each even sampling depth.

*qiime diversity alpha-rarefaction *
    *--i-table table.qza *
    *--i-phylogeny rooted-tree.qza *
    *--p-max-depth 4000 *

```
--m-metadata-file sample-metadata.tsv \
--o-visualization alpha-rarefaction.qzv
```

Output visualizations:

alpha-rarefaction.qzv

Taxonomic analysis

In this section, the taxonomic composition of the samples will be explored and related to the sample metadata. The first step in this process is to assign taxonomy to the sequences in our Feature Data [Sequence] QIIME 2 artifact. Using a pre-trained Naive Bayes classifier and the q2-feature-classifier plugin on the Greengenes 13_8 99% OTUs, only include 250 bases from the region of the 16S that was sequenced in this analysis (the V4 region, bound by the 515F/806R primer pair). The application of this classifier to training datasets generates a visualization of mapping, i.e., from sequence to taxonomy.

```
wget \
  -O "gg-13-8-99-515-806-nb-classifier.qza" \
  "https://data.qiime2.org/2024.2/common/gg-13-8-99-515-806-nb-classifier.qza"

qiime feature-classifier classify-sklearn \
  --i-classifier gg-13-8-99-515-806-nb-classifier.qza \
  --i-reads rep-seqs.qza \
  --o-classification taxonomy.qza

qiime metadata tabulate \
  --m-input-file taxonomy.qza \
  --o-visualization taxonomy.qzv
```

Create interactive bar plots of training samples with the following command and open the visualization to see their taxonomic composition.

```
qiime taxa barplot \
  --i-table table.qza \
  --i-taxonomy taxonomy.qza \
  --m-metadata-file sample-metadata.tsv \
  --o-visualization taxa-bar-plots.qzv
```

Differential abundance testing with ANCOM-BC

ANCOM-BC can be applied to identify features that are differentially abundant (i.e., present in different abundances) across sample groups. We recommend to review the assumptions and limitations of ANCOM-BC before using it (Lin and Peddada, 2020). The q2-composition plugin presently uses ANCOM-BC, a compositionally-aware linear regression model that enables testing differentially abundant features across groups and incorporates bias correction. Several features differ significantly between body sites. Therefore, to identify any sequence variants and genera that are differentially abundant between the gut samples of two subjects, use only gut samples and apply ANCOM-BC.

```
qiime feature-table filter-samples \
   --i-table table.qza \
   --m-metadata-file sample-metadata.tsv \
   --p-where "[body-site]='gut'" \
   --o-filtered-table gut-table.qza
```

Output artifacts:

gut-table.qza

ANCOM-BC operates on a Feature Table [Frequency] QIIME 2 artifact. Therefore, perform ANCOM-BC on the subject column to determine what features differ in abundance across gut samples of the two subjects.

```
qiime composition ancombc \
   --i-table gut-table.qza \
   --m-metadata-file sample-metadata.tsv \
   --p-formula 'subject' \
   --o-differentials ancombc-subject.qza
```

```
qiime composition da-barplot \
   --i-data ancombc-subject.qza \
   --p-significance-threshold 0.001 \
   --o-visualization da-barplot-subject.qzv
```

Output artifacts:

ancombc-subject.qza

Output visualizations:

da-barplot-subject.qzv

If interested in performing a differential abundance test at a specific taxonomic level, collapse the features in our Feature Table [Frequency] at the taxonomic level of interest (preferably at the genus level (i.e., level 6 of the Greengenes taxonomy)), and then re-run the above steps.

```
qiime taxa collapse \
   --i-table gut-table.qza \
   --i-taxonomy taxonomy.qza \
   --p-level 6 \
   --o-collapsed-table gut-table-l6.qza
```

```
qiime composition ancombc \
   --i-table gut-table-l6.qza \
   --m-metadata-file sample-metadata.tsv \
   --p-formula 'subject' \
   --o-differentials l6-ancombc-subject.qza

qiime composition da-barplot \
   --i-data l6-ancombc-subject.qza \
   --p-significance-threshold 0.001 \
   --p-level-delimiter ';' \
   --o-visualization l6-da-barplot-subject.qzv
```

Output artifacts:

```
gut-table-l6.qza
l6-ancombc-subject.qza
```

Output visualizations:

```
l6-da-barplot-subject.qzv
```

Supervised machine learning using QIIME2

q2-sample-classifier

The supervised machine learning techniques supported by this QIIME 2 plugin include techniques for sample metadata regression and classification. Microbiome investigations frequently use supervised learning techniques to effectively carry out tasks like result prediction and sample differentiation based on microbial composition. For an enormous number of microbiologists, clinicians, and others who want to use supervised learning methods for predicting sample characteristics based on microbiome composition or other "omics" data, the q2-sample-classifier plugin improves accessibility, reproducibility, and interpretability of these techniques (Pedregosa et al., 2011; Bokulich et al., 2018).

q2cli Invocation

qiime sample-classifier

Artifact API Import

from qiime2.plugins import sample_classifier

Acknowledgments

Disclaimer about adopted content from QIIME 2 website. All contents in this chapter are adopted as it is to avoid misunderstanding with developer and use of commands from (https://qiime2.org/ and https://docs.qiime2.org/2024.2/tutorials/moving-pictures/).

References

Anderson, M. J. (2017). Permutational multivariate analysis of variance (PERMANOVA). Wiley StatsRef: Statistics Reference Online. doi: 10.1002/9781118445112.stat07841.

Barr, T., Sureshchandra, S., Ruegger, P., Zhang, J., Ma, W., Borneman, J. et al. (2018). Concurrent gut transcriptome and microbiota profiling following chronic ethanol consumption in nonhuman primates. Gut Microbes 9(4): 338–356. https://doi.org/10.1080/19490976.2018.1441663.

Bokulich, N. A., Dillon, M. R., Bolyen, E., Kaehler, B. D., Huttley, G. A. and Caporaso, J. G. (2018). q2-sample-classifier: machine-learning tools for microbiome classification and regression. Journal of Open Research Software 3(30): 934. https://doi.org/10.21105/joss.00934.

Bokulich, N. A., Subramanian, S., Faith, J. J., Gevers, D., Gordon, J. I., Knight, R. et al. (2013). Quality-filtering vastly improves diversity estimates from Illumina amplicon sequencing. Nature Methods 10(1): 57–59. https://doi.org/10.1038/nmeth.2276.

Bolyen, E., Rideout, J. R., Dillon, M. R., Bokulich, N. A., Abnet, C. C., Al-Ghalith, G. A. et al. (2019). Reproducible, interactive, scalable and extensible microbiome data science using QIIME 2. Nature Biotechnology 37(8): 852–857. https://doi.org/10.1038/s41587-019-0209-9.

Caporaso, J. G., Kuczynski, J., Stombaugh, J., Bittinger, K., Bushman, F. D., Costello, E. K. et al. (2010). QIIME allows analysis of high-throughput community sequencing data. Nature Methods 7(5): 335–336. https://doi.org/10.1038/nmeth.f.303.

Caporaso, J. G., Kuczynski, J., Stombaugh, J., Bittinger, K., Bushman, F. D., Costello, E. K. et al. (2010). QIIME allows analysis of high-throughput community sequencing data. Nature Methods 7(5): 335–336. https://doi.org/10.1038/nmeth.f.303.

Gehring, C. A., Sthultz, C. M., Flores-Rentería, L., Whipple, A. V. and Whitham, T. G. (2017). Tree genetics defines fungal partner communities that may confer drought tolerance. Proceedings of the National Academy of Sciences of the United States of America 114(42): 11169–11174. https://doi.org/10.1073/pnas.1704022114.

Gopalakrishnan, V., Spencer, C. N., Nezi, L., Reuben, A., Andrews, M. C., Karpinets, T. V. et al. (2018). Gut microbiome modulates response to anti-PD-1 immunotherapy in melanoma patients. Science (New York, N.Y.) 359(6371): 97–103. https://doi.org/10.1126/science.aan4236.

Kapono, C. A., Morton, J. T., Bouslimani, A., Melnik, A. V., Orlinsky, K., Knaan, T. L. et al. (2018). Creating a 3D microbial and chemical snapshot of a human habitat. Scientific Reports 8(1): 3669. https://doi.org/10.1038/s41598-018-21541-4.

Lee, K., Pletcher, S. D., Lynch, S. V., Goldberg, A. N. and Cope, E. K. (2018). Heterogeneity of microbiota dysbiosis in chronic rhinosinusitis: potential clinical implications and microbial community mechanisms contributing to sinonasal inflammation. Frontiers in Cellular and Infection Microbiology 8: 168. https://doi.org/10.3389/fcimb.2018.00168.

Lin, H. and Peddada, S. D. (2020). Analysis of compositions of microbiomes with bias correction. Nature Communications 11: 3514. https://doi.org/10.1038/s41467-020-17041-7.

Metcalf, J. L., Xu, Z. Z., Weiss, S., Lax, S., Van Treuren, W., Hyde, E. R. et al. (2016). Microbial community assembly and metabolic function during mammalian corpse decomposition. Science (New York, N.Y.) 351(6269): 158–162. https://doi.org/10.1126/science.aad2646.

Pedregosa, F., Varoquaux, G., Gramfort, A., Michel, V., Thirion, B., Grisel, O. et al. (2011) Scikit-learn: machine learning in python. Journal of Machine Learning Research 12(Oct): 2825–2830.

Pineda, A., Kaplan, I. and Bezemer, T. M. (2017). Steering soil microbiomes to suppress aboveground insect pests. Trends in Plant Science 22(9): 770–778. https://doi.org/10.1016/j.tplants.2017.07.002.

Rekadwad, B. N., Shouche, Y. S. and Jangid, K. (2024). Community characterization of soil from the Central Deccan Plateau dry tropical deciduous forest (Ceddtrof) in central India using 16S rRNA gene amplicon sequencing. Microbiology Resource Announcements 13(2): e0113423. https://doi.org/10.1128/mra.01134-23.

Rubin, R. L., Koch, G. W., Martinez, A., Mau, R. L., Bowker, M. A. and Hungate, B. A. (2018). Developing climate-smart restoration: Can plant microbiomes be hardened against heat waves?. Ecological Applications : A Publication of the Ecological Society of America 28(6): 1594–1605. https://doi.org/10.1002/eap.1763.

Smith, M. I., Yatsunenko, T., Manary, M. J., Trehan, I., Mkakosya, R., Cheng, J. et al. (2013). Gut microbiomes of Malawian twin pairs discordant for kwashiorkor. Science (New York, N.Y.) 339(6119): 548–554. https://doi.org/10.1126/science.1229000.

Verberkmoes, N. C., Russell, A. L., Shah, M., Godzik, A., Rosenquist, M., Halfvarson, J. et al. (2009). Shotgun metaproteomics of the human distal gut microbiota. The ISME Journal 3(2): 179–189. https://doi.org/10.1038/ismej.2008.108.

Chapter 9
Microbial Dark Matter
Taxonomy, Importance, Ethical and Philosophical Considerations

Introduction

Investigations of microbial communities from various worldwide habitats have shown a wealth of new species and genes, as well as distinct spatiotemporal patterns amongst locales. However, there is still a significant quantity of microbial variety that has not yet been identified. What has been called "microbial dark matter" is a vast reservoir of diversity made up of these unique community structures and functions. Microbial dark matter is made up of many elements (Integrative HMP (iHMP) Research Network Consortium, 2019; Thompson et al., 2017; Sunagawa et al., 2015a; 2015b; Knights et al., 2011; Shenhav et al., 2019). Microbial communities live in millions of biomes, or niches, including broad settings like freshwaters and soils, context-dependent biomes, and little-studied biomes like the gut microbiomes of patients with various illnesses (Aggarwala et al., 2017; Carroll et al., 2018; Jonas and Seifman, 2019; Biteen et al., 2016; Liu et al., 2019; Bäckhed et al., 2015; Bashan et al., 2016; Smits et al., 2017). Furthermore, tens of millions of microbial species from other life kingdoms, such as bacteria, viruses, protists, and archaea, are known to exist. Moreover, the genomes of microbial communities encode billions of functional genes. Lastly, the compositions of microbial communities are influenced by an infinite number of dynamic ecological and evolutionary phenomena. Though many of these aspects of microbial dark matter are still understudied, they all hold significant potential for improving our understanding of the microbial world (Halfvarson et al., 2017; Turner et al., 2013; Surana and Kasper, 2017; Gilbert et al., 2012; Minot et al., 2011;

Virgin, 2014; Handley, 2016; Zha et al., 2022). The mining of a significant quantity of microbiome data can assist in the generation of information and models for a wide range of applications. These applications include the identification of new species and genes, the tracking of sample sources, the prediction of phenotypes (particularly for the diagnosis of diseases), and the development of prediction models for longitudinal studies (Zha et al., 2022).

Habitat of microbial dark matter

The realm of microbial dark matter extends far beyond our conventional understanding of habitable spaces (Orcutt et al., 2011; Solden et al., 2016; Escudeiro et al., 2022; Li et al., 2023; Schultz et al., 2023; Kumar et al., 2022). Within the human microbiome, the uncharted regions of our gut, skin, and various internal organs teem with uncultured microorganisms, hinting at their intricate roles in our health (Ogunrinola et al., 2020; Afzaal et al., 2022; Dekaboruah et al., 2020). Extreme environments, once thought hostile to life, are now hotbeds of microbial dark matter—from the crushing pressure of the deep ocean to the scorching heat of geothermal springs, and even in the seemingly barren depths of the subsurface. Soil, teeming with complexity, harbours a vast reservoir of microbial dark matter, essential to nutrient cycling and plant health (Schultz et al., 2013; Kalwasińska et al., 2020; Kostešić et al., 2023; Shu and Huang, 2022; Kochhar et al., 2022). The surfaces and internal tissues of plants form intimate partnerships with unique communities of microbes, much of which remains unclassified (Turner et al., 2013; Jacoby et al., 2017; Schlaeppi and Bulgarelli, 2015; Cordovez et al., 2019). And even more unexpectedly, microbial dark matter finds refuge in niches as diverse as ancient cave systems, contaminated industrial sites, and even the built environment around us (Wong et al., 2020; Kosznik-Kwaśnicka et al., 2022; Kapinusova et al., 2023).

Hidden potential of microbial dark matter

Fortunately, the pioneering approach, which was initially largely based on the development of 16-S rRNA gene sequencing (Schmidt et al., 1991; Barns et al., 1996; Hugenholtz et al., 1998), then on the sequencing of other manufacturers (Beja et al., 2000), and later on the development of metagenomics (Breitbart et al., 2002; Tyson et al., 2005; Tringe et al., 2005) and single-cell genomics, managed to circumvent the requirement for culture studies, thereby removing a blind spot that was imposed by culture-based investigation to comparative analyses. The results of these experiments revealed a wide range of fascinating discoveries. Microbial ecologists had begun to characterize the gene content, variety, and relative abundance of environmental microorganisms by the beginning of the 2000s (Venter et al., 2004). Additionally, they had begun,. new activities of considerable relevance in the ocean had been discovered by them (for example, the oxidation of ammonia by archaea; Francis et al., 2005). These functions might potentially have an effect on the global nitrogen cycle. Additionally, they discovered unexpected genes in viruses that are involved in photosynthesis and other processes (Sullivan et al., 2005). They had also gained insights into the survival strategies of microbes that had never been seen before (Tyson et al., 2004), into the community structures of microbes (Tyson et al., 2004; DeLong et al., 2006), and into the niche-specific adaptations of microbes (Tringe et al., 2005). For instance, they had discovered previously unknown iron-oxidizing and free-living diazotrophs in acid mine drainage biofilms (Ram et al., 2005; Tyson et al., 2005).

At the same time when new microorganisms were discovered, our understanding of the processes that involved or are affected by bacteria underwent a significant evolution. As a prominent trend, the emphasis on interactions and the utilization of networks rather than trees as a framework for microbial research is becoming increasingly prevalent (Cluster et al., 2022; Stencel and Wloch-Salamon, 2018; Bapteste et al., 2021; Ferraz Helene et al., 2022). The diversity and complexity of the processes that explain the evolution of microbes are not being adequately represented by simple tree-based models, which are attempting to recreate the divergence of lineages from a last common ancestor. This is becoming increasingly apparent. For instance, in the natural world, diversity-generating retroelements play a role in the rapid and targeted sequence diversification that occurs in Archaea and associated viruses (Paul et al., 2015), as well as in CPR (Paul et al., 2017). Doolittle (1999), Ochman et al. (2000), and Bapteste et al. (2012) all point out that the collaborative aspect of microbial evolution is particularly stressed by introgressive mechanisms such as lateral gene transfer. The importance of metabolic, ecological, and evolutionary scaffolding in the microbial world is further supported by the discovery of syntrophic consortia and environmental microbes with genuinely incomplete genomes (i.e., lacking genes deemed essential) (DeLong, 2007; Morris et al., 2012; Sachs and Hollowell, 2012; Caporael et al., 2013; Brown et al., 2015; Ereshefsky and Pedroso, 2015).

With the idea that natural selection ultimately promotes individual optimized lineages through the success of the fittest cells amid huge and phylogenetically homogeneous microbial populations, the claim that bacteria in nature are dependent on other organisms to survive stands in stark contrast to the idea that natural selection ultimately favours individual optimized lineages. On the other hand, it is in agreement with the actual observation that pure culture is ineffective for the majority of microorganisms (Staley and Konopka, 1985), and it does in reality provide an explanation for this significant plate anomaly. Microbes are not solitary organisms; rather, they are members of collective bodies. As researchers continue to delve deeper into the world of microbes, they are discovering a number of other fascinating relationships. As an illustration, the dynamics of microbial and viral populations are influenced by types of communication that are not heard (Erez et al., 2017). Some researchers propose the introduction of holobionts, which are the emergent associations of hosts and microbes, as a novel kind of central evolutionary player (Bordenstein and Theis, 2015; Moran and Sloan, 2015; Theis et al., 2016). This is due to the fact that microbiomes and their hosts are responsible for the construction of a wide variety of animal and plant phenotypes (Gill et al., 2006; Gilbert et al., 2015). On an even larger scale, it is now believed that microbes, the majority of which are unknown, have an effect on the geochemical processes that shape our planet (Guidi et al., 2016). Furthermore, through a process known as niche construction (Laland et al., 2016), these microbes are thought to have the potential to influence ecosystems and the future of life. Despite the fact that all of these processes—lateral gene transfer, scaffolding, communication, microbial co construction, and niche construction—are quite common in the world of microbes, they are still considered to be somewhat peripheral in terms of explaining biological phenomena. Therefore, in order to introduce the processes to which microbial dark matter contribute within biological theory, it is necessary to revise the relative priority that is now ascribed to concepts in scientific explanations. This is likely to be a protracted and labourious epistemic process. For instance, prokaryotic biology, particularly when microbiomes are taken into consideration, appears to be so dissimilar to the biology of model eukaryotic organisms. There have been a number of evolutionary biologists and theorists who have independently proposed that key aspects of the classic Darwinian theory and of the Modern Synthesis would have been very different if microbial studies had been more central during the early stages of the development of the evolutionary theory. On the other hand, there are many who disagree with the notion that the evolutionary theory needs to be reformed in terms of both its structure and its content, even in light of the newest discoveries in the field of microbiology (Laland et al., 2014).

Nonetheless, discussions regarding the makeup, phylogenetic position, and gene content of Asgard archaea (Saw et al., 2015; Da Cunha et al., 2017; Zaremba-Niedzwiedzka et al., 2017) clearly demonstrate how crucial components of evolutionary theory could change with a better understanding of microbial dark matter. In the event that Asgard archaea, which are currently only known through the assembly of environmental reads, are found to be sister-groups of eukaryotes, this should (at the very least) have an effect on the concept of a tree of life. It should also bring additional evidence regarding the number of domains of life (given that there is a convincing argument that the two domains tree is better supported than the three domains tree predates the discovery of Asgard; Williams et al., 2013). Furthermore, depending on the intimate structural biology and metabolisms of these Asgard, it will also assist in testing among competing hypotheses for the origin of eukaryotes (Koonin, 2015; Sousa et al., 2016).

On a different level, the recently found genes of microorganisms have also had an impact, and they may continue to have an impact, on essential social requirements. The development of the industrial enzymes market, which was projected to reach up to US dollars 6.20 billion in 2020, depended on the discovery of new antibiotics in the environment (Lok, 2015), such as Teixobactin (Ling et al., 2015), or enzymes with greater activity, specificity, or stability, such as lipases (Rogalska et al., 1997) or organo-phosphorus degrading enzymes (Singh, 2009). Research in the scientific community has also benefited tremendously from the discovery of microbial enzymes, which has been recognized by a number of Nobel Prizes. These enzymes include restrictions enzymes, such as HindII (Smith and Wilcox, 1970), or DNA polymerases (Brock and Freeze, 1969), which made it possible to develop the Polymerase Chain Reaction (Saiki et al., 1988). The discovery of Crispr-cas9 systems (Jinek et al., 2012), which are now employed for genome editing, further demonstrates the great potential of microbial gene discovery to promote the development of pharmaceuticals, biotechnologies, and research tools. This discovery was made more recently (Bernard et al., 2018).

Taxonomy of microbial dark matter

Reconstructing the full evolutionary history of living life on Earth will be a major scientific achievement, similar to the periodic table, which revolutionized chemistry. Genomes are our most complete and objective evolutionary documents, therefore comparative genomics leads to this goal. Over the past decade, plant and animal genomes have been targeted to cover the tree of life. Multicellular organisms only appeared in the last 550 million years of biological history, making up a minuscule fraction of biological variety. The majority of past and contemporary biodiversity is microbial. Most microbes cannot be produced in the laboratory and are unknown to science, therefore we know little about them. Over the past 30 years, advanced culture-independent molecular techniques have opened the door to microbial dark matter, which is accelerating due to exponential sequencing capacity. First-ever representative genomes of all life are imminent. The highly incomplete tree is filled with taxonomic mistakes due to past usage of morphology, biochemical properties, behavioural features, and single-marker genes to infer organismal connections. Creating a coherent global tree of life and genome-based taxonomy from growing genomic data sets requires concerted work (Hugenholtz et al., 2016).

Typically, groupings of "unknown microorganisms" resulting from misaligned sequences are disregarded and excluded from diversity analyses. We examined the 16S rRNA gene sequences of microbial communities from four distinct environments—a living thing, a desert, a naturally occurring aquatic habitat, and a membrane bioreactor for wastewater treatment—in order to close this information gap. "Microbial dark matter sequences" (MDMS) are typical sequences of potentially unknown microorganisms selected from those databases for further analysis. By using targeted amplification and resequencing, the existence of the sequence was confirmed. To create a thorough phylogenetic tree for further sequence categorization, these sequences were compared to databases and aligned using the Genome Taxonomy Database. This process revealed perhaps new candidate phyla and other lineages. In order to provide further confirmation and to characterize the taxonomic and metabolic aspects of these potential MDMS, they were further compared to metagenome-assembled genomes from the investigated habitats. This chapter provides a straightforward and easily accessible methodology for the investigation of MDM concealed behind amplicon sequencing findings, demonstrating the enormous significance of MDMS in environmental metataxonomic analysis of 16S rRNA gene sequences (Méheust et al., 2019; Pascoal et al., 2021; Zamkovaya et al., 2021; Wiegand et al., 2021; Mise and Iwaski, 2022; Barak et al., 2023).

Ethical and philosophical considerations of microbial dark matter sequence use for taxonomy and identification

An ethical perspective influences technological advancements, which are influenced by the technology that is now accessible. There are many advances in biotechnology that are good for humanity. But there is a darker side to technology. Unexpected effects of biotechnology have the potential to damage or dehumanize individuals. It is necessary to carefully consider the ethical consequences of proposed advancements (O'Mathúna, 2007). The cultivation-independent sequencing methods' genomic data continue to be the main source of understanding the Earth's uncultivated microbiome, or "microbial dark matter" (MDM) (Bernard et al., 2018; Brown et al., 2015; Dam et al., 2020; Rinke et al., 2013; Kaster and Sobol, 2020; Pratscher et al., 2018). The current MAGs and SAGs that have been publically deposited suggest the presence of contaminated contigs in a considerable number of genomes that are considered to be of "high quality." Some of these genomes even passed the quality checks conducted by GTDB and were considered to be representative genomes for the purposes of reference databases. The implementation of an alternate method for the detection and filtering of potential contaminants in Metgenome-Associated Genomes (MAGs) and Single-cell Amplified Genomes (SAGs) that are incomplete or fragmented. Individual researchers using the pipeline on different new genomes can effectively find contaminated reference databases using this method. This allows for ongoing maintenance of the reference database and stops errors from spreading and corrupting it over time. This will ensure that our picture of microbial dark matter is maintained in a manner that is progressively obvious with each new MAG or SAG that is analyzed and submitted. Additionally, it will prevent lingering contaminants in submitted genomes from gradually obscuring our perspective of dark matter (Vollmers et al., 2022).

References

Afzaal, M., Saeed, F., Shah, Y. A., Hussain, M., Rabail, R., Socol, C. T. et al. (2022). Human gut microbiota in health and disease: Unveiling the relationship. Frontiers in Microbiology 13: 999001. https://doi.org/10.3389/fmicb.2022.999001.

Aggarwala, V., Liang, G. and Bushman, F. D. (2017). Viral communities of the human gut: metagenomic analysis of composition and dynamics. Mobile DNA 8: 12. https://doi.org/10.1186/s13100-017-0095-y.

Bäckhed, F., Roswall, J., Peng, Y., Feng, Q., Jia, H., Kovatcheva-Datchary, P. et al. (2015). Dynamics and stabilization of the human gut microbiome during the first year of life. Cell Host & Microbe 17(6): 852. https://doi.org/10.1016/j.chom.2015.05.012.

Bapteste, E., Lopez, P., Bouchard, F., Baquero, F., McInerney, J. O. and Burian, R. M. (2012). Evolutionary analyses of non-genealogical bonds produced by introgressive descent. Proceedings of the National Academy of Sciences of the United States of America 109(45): 18266–18272. https://doi.org/10.1073/pnas.1206541109.

Bapteste, E., Gérard, P., Larose, C., Blouin, M., Not, F., Campos, L. et al. (2021). The epistemic revolution induced by microbiome studies: an interdisciplinary view. Biology 10(7): 651. https://doi.org/10.3390/biology10070651.

Barak, H., Fuchs, N., Liddor-Naim, M., Nir, I., Sivan, A. and Kushmaro, A. (2023). Microbial dark matter sequences verification in amplicon sequencing and environmental metagenomics data. Frontiers in Microbiology 14: 1247119. https://doi.org/10.3389/fmicb.2023.1247119.

Barns, S. M., Delwiche, C. F., Palmer, J. D. and Pace, N. R. (1996). Perspectives on archaeal diversity, thermophily and monophyly from environmental rRNA sequences. Proceedings of the National Academy of Sciences of the United States of America 93(17): 9188–9193. https://doi.org/10.1073/pnas.93.17.9188.

Bashan, A., Gibson, T. E., Friedman, J., Carey, V. J., Weiss, S. T., Hohmann, E. L. et al. (2016). Universality of human microbial dynamics. Nature 534(7606): 259–262. https://doi.org/10.1038/nature18301.

Béjà, O., Aravind, L., Koonin, E. V., Suzuki, M. T., Hadd, A., Nguyen, L. P. et al. (2000). Bacterial rhodopsin: evidence for a new type of phototrophy in the sea. Science (New York, N.Y.) 289(5486): 1902–1906. https://doi.org/10.1126/science.289.5486.1902.

Bernard, G., Pathmanathan, J. S., Lannes, R., Lopez, P. and Bapteste, E. (2018). Microbial dark matter investigations: how microbial studies transform biological knowledge and empirically sketch a logic of scientific discovery. Genome Biology and Evolution 10(3): 707–715. https://doi.org/10.1093/gbe/evy031.

Biteen, J. S., Blainey, P. C., Cardon, Z. G., Chun, M., Church, G. M., Dorrestein, P. C. et al. (2016). Tools for the microbiome: nano and beyond. ACS Nano 10(1): 6–37. https://doi.org/10.1021/acsnano.5b07826.

Bordenstein, S. R. and Theis, K. R. (2015). Host biology in light of the microbiome: ten principles of holobionts and hologenomes. PLoS Biology 13(8): e1002226. https://doi.org/10.1371/journal.pbio.1002226.

Breitbart, M., Salamon, P., Andresen, B., Mahaffy, J. M., Segall, A. M., Mead, D. et al. (2002). Genomic analysis of uncultured marine viral communities. Proceedings of the National Academy of Sciences of the United States of America 99(22): 14250–14255. https://doi.org/10.1073/pnas.202488399.

Brock, T. D. and Freeze, H. (1969). Thermus aquaticus gen. n. and sp. n., a nonsporulating extreme thermophile. Journal of Bacteriology 98(1): 289–297. https://doi.org/10.1128/jb.98.1.289-297.1969.

Brown, C. T., Hug, L. A., Thomas, B. C., Sharon, I., Castelle, C. J., Singh, A. et al. (2015). Unusual biology across a group comprising more than 15% of domain Bacteria. Nature 523(7559): 208–211. https://doi.org/10.1038/nature14486.

Bull, M. J. and Plummer, N. T. (2014). Part 1: The human gut microbiome in health and disease. Integrative Medicine (Encinitas, Calif.) 13(6): 17–22.

Caporael, L., Griesemer, J. and Wimsatt, W. (2013). Scaffolding in Evolution, Culture, and Cognition. Massachusetts: MIT Press.

Carroll, D., Daszak, P., Wolfe, N. D., Gao, G. F., Morel, C. M., Morzaria, S. et al. (2018). The Global Virome Project. Science (New York, N.Y.) 359(6378): 872–874. https://doi.org/10.1126/science.aap7463.

Cordovez, V., Dini-Andreote, F., Carrión, V. J. and Raaijmakers, J. M. (2019). Ecology and evolution of plant microbiomes. Annual Review of Microbiology 73: 69–88. https://doi.org/10.1146/annurev-micro-090817-062524.

Custer, G. F., Bresciani, L. and Dini-Andreote, F. (2022). Ecological and evolutionary implications of microbial dispersal. Frontiers in Microbiology 13: 855859. https://doi.org/10.3389/fmicb.2022.855859.

Da Cunha, V., Gaia, M., Gadelle, D., Nasir, A. and Forterre, P. (2017). Lokiarchaea are close relatives of Euryarchaeota, not bridging the gap between prokaryotes and eukaryotes. PLoS Genetics 13(6): e1006810. https://doi.org/10.1371/journal.pgen.1006810.

Dam, H. T., Vollmers, J., Sobol, M. S., Cabezas, A. and Kaster, A. K. (2020). Targeted cell sorting combined with single cell genomics captures low abundant microbial dark matter with higher sensitivity than metagenomics. Frontiers in Microbiology 11: 1377. https://doi.org/10.3389/fmicb.2020.01377.

de Vos, W. M., Tilg, H., Van Hul, M. and Cani, P. D. (2022). Gut microbiome and health: mechanistic insights. Gut 71(5): 1020–1032. https://doi.org/10.1136/gutjnl-2021-326789.

Dekaboruah, E., Suryavanshi, M. V., Chettri, D. and Verma, A. K. (2020). Human microbiome: an academic update on human body site specific surveillance and its possible role. Archives of Microbiology 202(8): 2147–2167. https://doi.org/10.1007/s00203-020-01931-x.

DeLong, E. F., Preston, C. M., Mincer, T., Rich, V., Hallam, S. J., Frigaard, N. U. et al. (2006). Community genomics among stratified microbial assemblages in the ocean's interior. Science (New York, N.Y.) 311(5760): 496–503. https://doi.org/10.1126/science.1120250.

DeLong, E. F. (2007). Microbiology. Life on the thermodynamic edge. Science (New York, N.Y.) 317(5836): 327–328. https://doi.org/10.1126/science.1145970.

Doolittle, W. F. (1999). Phylogenetic classification and the universal tree. Science (New York, N.Y.) 284(5423): 2124–2129. https://doi.org/10.1126/science.284.5423.2124.

Ereshefsky, M. and Pedroso, M. (2015). Rethinking evolutionary individuality. Proceedings of the National Academy of Sciences of the United States of America 112(33): 10126–10132. https://doi.org/10.1073/pnas.1421377112.

Erez, Z., Steinberger-Levy, I., Shamir, M., Doron, S., Stokar-Avihail, A., Peleg, Y. et al. (2017). Communication between viruses guides lysis-lysogeny decisions. Nature 541(7638): 488–493. https://doi.org/10.1038/nature21049.

Escudeiro, P., Henry, C. S. and Dias, R. P. M. (2022). Functional characterization of prokaryotic dark matter: the road so far and what lies ahead. Current Research in Microbial Sciences 3: 100159. https://doi.org/10.1016/j.crmicr.2022.100159.

Ferraz Helene, L. C., Klepa, M. S. and Hungria, M. (2022). New insights into the taxonomy of bacteria in the genomic era and a case study with rhizobia. International Journal of Microbiology 2022: 4623713. https://doi.org/10.1155/2022/4623713.

Francis, C. A., Roberts, K. J., Beman, J. M., Santoro, A. E. and Oakley, B. B. (2005). Ubiquity and diversity of ammonia-oxidizing archaea in water columns and sediments of the ocean. Proceedings of the National Academy of Sciences of the United States of America 102(41): 14683–14688. https://doi.org/10.1073/pnas.0506625102.

Gilbert, J. A., Steele, J. A., Caporaso, J. G., Steinbrück, L., Reeder, J., Temperton, B. et al. (2012). Defining seasonal marine microbial community dynamics. The ISME Journal 6(2): 298–308. https://doi.org/10.1038/ismej.2011.107.

Gilbert, S. F., Bosch, T. C. and Ledón-Rettig, C. (2015). Eco-Evo-Devo: developmental symbiosis and developmental plasticity as evolutionary agents. Nature Reviews. Genetics 16(10): 611–622. https://doi.org/10.1038/nrg3982.

Gill, S. R., Pop, M., Deboy, R. T., Eckburg, P. B., Turnbaugh, P. J., Samuel, B. S. et al. (2006). Metagenomic analysis of the human distal gut microbiome. Science (New York, N.Y.) 312(5778): 1355–1359. https://doi.org/10.1126/science.1124234.

Guidi, L., Chaffron, S., Bittner, L., Eveillard, D., Larhlimi, A., Roux, S. et al. (2016). Plankton networks driving carbon export in the oligotrophic ocean. Nature 532(7600): 465–470. https://doi.org/10.1038/nature16942.

Halfvarson, J., Brislawn, C. J., Lamendella, R., Vázquez-Baeza, Y., Walters, W. A., Bramer, L. M. et al. (2017). Dynamics of the human gut microbiome in inflammatory bowel disease. Nature Microbiology 2: 17004. https://doi.org/10.1038/nmicrobiol.2017.4.

Handley, S. A. (2016). The virome: a missing component of biological interaction networks in health and disease. Genome Medicine 8(1): 32. https://doi.org/10.1186/s13073-016-0287-y.

Hugenholtz, P., Goebel, B. M. and Pace, N. R. (1998). Impact of culture-independent studies on the emerging phylogenetic view of bacterial diversity. Journal of Bacteriology 180(18): 4765–4774. https://doi.org/10.1128/JB.180.18.4765-4774.1998.

Hugenholtz, P., Skarshewski, A. and Parks, D. H. (2016). Genome-based microbial taxonomy coming of age. Cold Spring Harbor Perspectives in Biology 8(6): a018085. https://doi.org/10.1101/cshperspect.a018085.

Human Microbiome Jumpstart Reference Strains Consortium, Nelson, K. E., Weinstock, G. M., Highlander, S. K., Worley, K. C., Creasy, H. H., Wortman, J. R. et al. (2010). A catalog of reference genomes from the human microbiome. Science (New York, N.Y.) 328(5981): 994–999. https://doi.org/10.1126/science.1183605.

Integrative HMP (iHMP) Research Network Consortium. (2019). The Integrative Human Microbiome Project. Nature 569(7758): 641–648. https://doi.org/10.1038/s41586-019-1238-8.

Jacoby, R., Peukert, M., Succurro, A., Koprivova, A. and Kopriva, S. (2017). The role of soil microorganisms in plant mineral nutrition-current knowledge and future directions. Frontiers in Plant Science 8: 1617. https://doi.org/10.3389/fpls.2017.01617.

Jinek, M., Chylinski, K., Fonfara, I., Hauer, M., Doudna, J. A. and Charpentier, E. (2012). A programmable dual-RNA-guided DNA endonuclease in adaptive bacterial immunity. Science (New York, N.Y.) 337(6096): 816–821. https://doi.org/10.1126/science.1225829.

Jonas, O. and Seifman, R. (2019). Do we need a Global Virome Project?. The Lancet. Global Health 7(10): e1314–e1316. https://doi.org/10.1016/S2214-109X(19)30335-3.

Kalwasińska, A., Krawiec, A., Deja-Sikora, E., Gołębiewski, M., Kosobucki, P., Swiontek Brzezinska, M. et al. (2020). Microbial diversity in deep-subsurface hot brines of northwest poland: from community structure to isolate characteristics. Applied and Environmental Microbiology 86(10): e00252-20. https://doi.org/10.1128/AEM.00252-20.

Kapinusova, G., Lopez Marin, M. A. and Uhlik, O. (2023). Reaching unreachables: Obstacles and successes of microbial cultivation and their reasons. Frontiers in Microbiology 14: 1089630. https://doi.org/10.3389/fmicb.2023.1089630.

Kaster, A. K. and Sobol, M. S. (2020). Microbial single-cell omics: the crux of the matter. Applied Microbiology and Biotechnology 104(19): 8209–8220. https://doi.org/10.1007/s00253-020-10844-0.

Knights, D., Kuczynski, J., Charlson, E. S., Zaneveld, J., Mozer, M. C., Collman, R. G. et al. (2011). Bayesian community-wide culture-independent microbial source tracking. Nature Methods 8(9): 761–763. https://doi.org/10.1038/nmeth.1650.

Kochhar, N., I K, K., Shrivastava, S., Ghosh, A., Rawat, V. S., Sodhi, K. K. and Kumar, M. (2022). Perspectives on the microorganism of extreme environments and their applications. Current Research in Microbial Sciences 3: 100134. https://doi.org/10.1016/j.crmicr.2022.100134.

Koonin, E. V. (2015). Archaeal ancestors of eukaryotes: not so elusive any more. BMC Biology 13: 84. https://doi.org/10.1186/s12915-015-0194-5.

Kostešić, E., Mitrović, M., Kajan, K., Marković, T., Hausmann, B., Orlić, S. et al. (2023). Microbial diversity and activity of biofilms from geothermal springs in croatia. Microbial Ecology 86(4): 2305–2319. https://doi.org/10.1007/s00248-023-02239-1.

Kosznik-Kwaśnicka, K., Golec, P., Jaroszewicz, W., Lubomska, D. and Piechowicz, L. (2022). Into the unknown: microbial communities in caves, their role, and potential use. Microorganisms 10(2): 222. https://doi.org/10.3390/microorganisms10020222.

Kumar, R., Sood, U., Kaur, J., Anand, S., Gupta, V., Patil, K. S. et al. (2022). The rising dominance of microbiology: what to expect in the next 15 years?. Microbial Biotechnology 15(1): 110–128. https://doi.org/10.1111/1751-7915.13953.

Laland, K., Uller, T., Feldman, M., Sterelny, K., Müller, G. B., Moczek, A. et al. (2014). Does evolutionary theory need a rethink?. Nature 514(7521): 161–164. https://doi.org/10.1038/514161a.

Laland, K., Matthews, B. and Feldman, M. W. (2016). An introduction to niche construction theory. Evolutionary Ecology 30: 191–202. https://doi.org/10.1007/s10682-016-9821-z.

Li, S., Lian, W. H., Han, J. R., Ali, M., Lin, Z. L., Liu, Y. H. et al. (2023). Capturing the microbial dark matter in desert soils using culturomics-based metagenomics and high-resolution analysis. NPJ Biofilms and Microbiomes 9(1): 67. https://doi.org/10.1038/s41522-023-00439-8.

Ling, L. L., Schneider, T., Peoples, A. J., Spoering, A. L., Engels, I., Conlon, B. P. et al. (2015). A new antibiotic kills pathogens without detectable resistance. Nature 517(7535): 455–459. https://doi.org/10.1038/nature14098.

Liu, H., Han, M., Li, S. C., Tan, G., Sun, S., Hu, Z. et al. (2019). Resilience of human gut microbial communities for the long stay with multiple dietary shifts. Gut 68(12): 2254–2255. https://doi.org/10.1136/gutjnl-2018-317298.

Lok, C. (2015). Mining the microbial dark matter. Nature 522(7556): 270–273. https://doi.org/10.1038/522270a.

Méheust, R., Burstein, D., Castelle, C. J. and Banfield, J. F. (2019). The distinction of CPR bacteria from other bacteria based on protein family content. Nature Communications 10(1): 4173. https://doi.org/10.1038/s41467-019-12171-z.

Minot, S., Sinha, R., Chen, J., Li, H., Keilbaugh, S. A., Wu, G. D. et al. (2011). The human gut virome: inter-individual variation and dynamic response to diet. Genome Research 21(10): 1616–1625. https://doi.org/10.1101/gr.122705.111.

Mise, K. and Iwasaki, W. (2022). Unexpected absence of ribosomal protein genes from metagenome-assembled genomes. ISME Communications 2(1): 118. https://doi.org/10.1038/s43705-022-00204-6.

Moran, N. A. and Sloan, D. B. (2015). The hologenome concept: helpful or hollow?. PLoS Biology 13(12): e1002311. https://doi.org/10.1371/journal.pbio.1002311.

Morris, J. J., Lenski, R. E. and Zinser, E. R. (2012). The Black Queen Hypothesis: evolution of dependencies through adaptive gene loss. mBio 3(2): e00036-12. https://doi.org/10.1128/mBio.00036-12.

O'Mathúna, D. P. (2007). Bioethics and biotechnology. Cytotechnology 53(1-3): 113–119. https://doi.org/10.1007/s10616-007-9053-8.

Ochman, H., Lawrence, J. G. and Groisman, E. A. (2000). Lateral gene transfer and the nature of bacterial innovation. Nature 405(6784): 299–304. https://doi.org/10.1038/35012500.

Ogunrinola, G. A., Oyewale, J. O., Oshamika, O. O. and Olasehinde, G. I. (2020). The human microbiome and its impacts on health. International Journal of Microbiology 2020: 8045646. https://doi.org/10.1155/2020/8045646.

Orcutt, B. N., Sylvan, J. B., Knab, N. J. and Edwards, K. J. (2011). Microbial ecology of the dark ocean above, at, and below the seafloor. Microbiology and Molecular Biology Reviews : MMBR 75(2): 361–422. https://doi.org/10.1128/MMBR.00039-10.

Pascoal, F., Costa, R. and Magalhães, C. (2021). The microbial rare biosphere: current concepts, methods and ecological principles. FEMS Microbiology Ecology 97(1): fiaa227. https://doi.org/10.1093/femsec/fiaa227.

Paul, B. G., Bagby, S. C., Czornyj, E., Arambula, D., Handa, S., Sczyrba, A. et al. (2015). Targeted diversity generation by intraterrestrial archaea and archaeal viruses. Nature Communications 6: 6585. https://doi.org/10.1038/ncomms7585.

Paul, B. G., Burstein, D., Castelle, C. J., Handa, S., Arambula, D., Czornyj, E. et al. (2017). Retroelement-guided protein diversification abounds in vast lineages of Bacteria and Archaea. Nature Microbiology 2: 17045. https://doi.org/10.1038/nmicrobiol.2017.45.

Pratscher, J., Vollmers, J., Wiegand, S., Dumont, M. G. and Kaster, A. K. (2018). Unravelling the identity, metabolic potential and global biogeography of the atmospheric methane-oxidizing upland soil cluster α. Environmental Microbiology 20(3): 1016–1029. https://doi.org/10.1111/1462-2920.14036.

Qin, J., Li, R., Raes, J., Arumugam, M., Burgdorf, K. S., Manichanh, C. et al. (2010). A human gut microbial gene catalogue established by metagenomic sequencing. Nature 464(7285): 59–65. https://doi.org/10.1038/nature08821.

Ram, R. J., Verberkmoes, N. C., Thelen, M. P., Tyson, G. W., Baker, B. J., Blake, R. C. et al. (2005). Community proteomics of a natural microbial biofilm. Science (New York, N.Y.) 308(5730): 1915–1920.

Rinke, C., Schwientek, P., Sczyrba, A., Ivanova, N. N., Anderson, I. J., Cheng, J. F. et al. (2013). Insights into the phylogeny and coding potential of microbial dark matter. Nature 499(7459): 431–437. https://doi.org/10.1038/nature12352.

Rogalska, E., Douchet, I. and Verger, R. (1997). Microbial lipases: structures, function and industrial applications. Biochemical Society Transactions 25(1): 161–164. https://doi.org/10.1042/bst0250161.

Sachs, J. L. and Hollowell, A. C. (2012). The origins of cooperative bacterial communities. mBio 3(3): e00099-12. https://doi.org/10.1128/mBio.00099-12.

Saiki, R. K., Gelfand, D. H., Stoffel, S., Scharf, S. J., Higuchi, R., Horn, G. T. et al. (1988). Primer-directed enzymatic amplification of DNA with a thermostable DNA polymerase. Science (New York, N.Y.) 239(4839): 487–491. https://doi.org/10.1126/science.2448875.

Saw, J. H., Spang, A., Zaremba-Niedzwiedzka, K., Juzokaite, L., Dodsworth, J. A., Murugapiran, S. K. et al. (2015). Exploring microbial dark matter to resolve the deep archaeal ancestry of eukaryotes. Philosophical Transactions of the Royal Society of London. Series B, Biological Sciences 370(1678): 20140328. https://doi.org/10.1098/rstb.2014.0328.

Schlaeppi, K. and Bulgarelli, D. (2015). The plant microbiome at work. Molecular Plant-Microbe Interactions : MPMI 28(3): 212–217. https://doi.org/10.1094/MPMI-10-14-0334-FI.

Schmidt, T. M., DeLong, E. F. and Pace, N. R. (1991). Analysis of a marine picoplankton community by 16S rRNA gene cloning and sequencing. Journal of Bacteriology 173(14): 4371–4378. https://doi.org/10.1128/jb.173.14.4371-4378.1991.

Schultz, J., Modolon, F., Peixoto, R. S. and Rosado, A. S. (2023). Shedding light on the composition of extreme microbial dark matter: alternative approaches for culturing extremophiles. Frontiers in Microbiology 14: 1167718. https://doi.org/10.3389/fmicb.2023.1167718.

Shenhav, L., Thompson, M., Joseph, T. A., Briscoe, L., Furman, O., Bogumil, D. et al. (2019). FEAST: fast expectation-maximization for microbial source tracking. Nature Methods 16(7): 627–632. https://doi.org/10.1038/s41592-019-0431-x.

Shu, W. S. and Huang, L. N. (2022). Microbial diversity in extreme environments. Nature reviews. Microbiology 20(4): 219–235. https://doi.org/10.1038/s41579-021-00648-y.

Singh, B. K. (2009). Organophosphorus-degrading bacteria: ecology and industrial applications. Nature reviews. Microbiology 7(2): 156–164. https://doi.org/10.1038/nrmicro2050.

Smith, H. O. and Wilcox, K. W. (1970). A restriction enzyme from Hemophilus influenzae. I. Purification and general properties. Journal of Molecular Biology 51(2): 379–391. https://doi.org/10.1016/0022-2836(70)90149-x.

Smits, S. A., Leach, J., Sonnenburg, E. D., Gonzalez, C. G., Lichtman, J. S., Reid, G. et al. (2017). Seasonal cycling in the gut microbiome of the Hadza hunter-gatherers of Tanzania. Science (New York, N.Y.) 357(6353): 802–806. https://doi.org/10.1126/science.aan4834.

Solden, L., Lloyd, K. and Wrighton, K. (2016). The bright side of microbial dark matter: lessons learned from the uncultivated majority. Curr. Opin. Microbiol. 2016 Jun;31:217-226. doi: 10.1016/j.mib.2016.04.020. Epub 2016 May 16. PMID: 27196505.

Sousa, F. L., Neukirchen, S., Allen, J. F., Lane, N. and Martin, W. F. (2016). Lokiarchaeon is hydrogen dependent. Nature Microbiology 1: 16034. https://doi.org/10.1038/nmicrobiol.2016.34.

Staley, J. T. and Konopka, A. (1985). Measurement of *in situ* activities of nonphotosynthetic microorganisms in aquatic and terrestrial habitats. Annual Review of Microbiology 39: 321–346. https://doi.org/10.1146/annurev.mi.39.100185.001541.

Stencel, A. and Wloch-Salamon, D. M. (2018). Some theoretical insights into the hologenome theory of evolution and the role of microbes in speciation. Theory in Biosciences = Theorie in den Biowissenschaften 137(2): 197–206. https://doi.org/10.1007/s12064-018-0268-3.

Sullivan, M. B., Coleman, M. L., Weigele, P., Rohwer, F. and Chisholm, S. W. (2005). Three Prochlorococcus cyanophage genomes: signature features and ecological interpretations. PLoS Biology 3(5): e144. https://doi.org/10.1371/journal.pbio.0030144.

Sunagawa, S., Coelho, L. P., Chaffron, S., Kultima, J. R., Labadie, K., Salazar, G. et al. (2015a). Ocean plankton. Structure and function of the global ocean microbiome. Science (New York, N.Y.) 348(6237): 1261359. https://doi.org/10.1126/science.1261359.

Sunagawa, S., Coelho, L. P., Chaffron, S., Kultima, J. R., Labadie, K., Salazar, G. et al. (2015b). Ocean plankton. Structure and function of the global ocean microbiome. Science (New York, N.Y.) 348(6237): 1261359. https://doi.org/10.1126/science.1261359.

Surana, N. K. and Kasper, D. L. (2017). Moving beyond microbiome-wide associations to causal microbe identification. Nature 552(7684): 244–247. https://doi.org/10.1038/nature25019.

Theis, K. R., Dheilly, N. M., Klassen, J. L., Brucker, R. M., Baines, J. F., Bosch, T. C. et al. (2016). Getting the hologenome concept right: an eco-evolutionary framework for hosts and their microbiomes. mSystems 1(2): e00028-16. https://doi.org/10.1128/mSystems.00028-16.

Thompson, L. R., Sanders, J. G., McDonald, D., Amir, A., Ladau, J., Locey, K. J. et al., Earth Microbiome Project Consortium. (2017). A communal catalogue reveals Earth's multiscale microbial diversity. Nature 551(7681): 457–463. https://doi.org/10.1038/nature24621.

Tringe, S. G., von Mering, C., Kobayashi, A., Salamov, A. A., Chen, K., Chang, H. W. et al. (2005). Comparative metagenomics of microbial communities. Science (New York, N.Y.) 308(5721): 554–557. https://doi.org/10.1126/science.1107851.

Turner, T. R., James, E. K. and Poole, P. S. (2013). The plant microbiome. Genome Biology 14(6): 209. https://doi.org/10.1186/gb-2013-14-6-209.

Tyson, G. W., Lo, I., Baker, B. J., Allen, E. E., Hugenholtz, P. and Banfield, J. F. (2005). Genome-directed isolation of the key nitrogen fixer Leptospirillum ferrodiazotrophum sp. nov. from an acidophilic microbial community. Applied and Environmental Microbiology 71(10): 6319–6324. https://doi.org/10.1128/AEM.71.10.6319-6324.2005.

Venter, J. C., Remington, K., Heidelberg, J. F., Halpern, A. L., Rusch, D., Eisen, J. A. et al. (2004). Environmental genome shotgun sequencing of the Sargasso Sea. Science (New York, N.Y.) 304(5667): 66–74. https://doi.org/10.1126/science.1093857.

Virgin, H. W. (2014). The virome in mammalian physiology and disease. Cell 157(1): 142–150. https://doi.org/10.1016/j.cell.2014.02.032.

Vollmers, J., Wiegand, S., Lenk, F. and Kaster, A. K. (2022). How clear is our current view on microbial dark matter? (Re-)assessing public MAG & SAG datasets with MDMcleaner. Nucleic Acids Research 50(13): e76. https://doi.org/10.1093/nar/gkac294.

Wiegand, S., Dam, H. T., Riba, J., Vollmers, J. and Kaster, A. K. (2021). Printing microbial dark matter: using single cell dispensing and genomics to investigate the patescibacteria/candidate phyla radiation. Frontiers in Microbiology 12: 635506. https://doi.org/10.3389/fmicb.2021.635506.

Williams, T. A., Foster, P. G., Cox, C. J. and Embley, T. M. (2013). An archaeal origin of eukaryotes supports only two primary domains of life. Nature 504(7479): 231–236. https://doi.org/10.1038/nature12779.

Wong, H. L., MacLeod, F. I., White, R. A., 3rd, Visscher, P. T. and Burns, B. P. (2020). Microbial dark matter filling the niche in hypersaline microbial mats. Microbiome 8(1): 135. https://doi.org/10.1186/s40168-020-00910-0.

Zamkovaya, T., Foster, J. S., de Crécy-Lagard, V. and Conesa, A. (2021). A network approach to elucidate and prioritize microbial dark matter in microbial communities. The ISME Journal 15(1): 228–244. https://doi.org/10.1038/s41396-020-00777-x.

Zaremba-Niedzwiedzka, K., Caceres, E. F., Saw, J. H., Bäckström, D., Juzokaite, L., Vancaester, E. et al. (2017). Asgard archaea illuminate the origin of eukaryotic cellular complexity. Nature 541(7637): 353–358. https://doi.org/10.1038/nature21031.

Zha, Y., Chong, H., Yang, P. and Ning, K. (2022). Microbial dark matter: from discovery to applications. Genomics, Proteomics & Bioinformatics 20(5): 867–881. https://doi.org/10.1016/j.gpb.2022.02.007.

Chapter 10
Future Horizons of Microbe Hunting
Unveiling the Secrets of the 22nd Century and Beyond

Introduction

The 21st century has witnessed a revolution in microbe hunting, fuelled by advancements like metagenomics and single-cell analysis. But the journey is far from over. With an estimated 10^{30} microbial cells on Earth, only a tiny fraction have been encountered, leaving the 22nd century brimming with exciting possibilities (Handelsman, 2004; Culligan et al., 2014). This vast unexplored environment holds immense potential for unlocking secrets of microbial life, driving innovation in diverse fields, and shaping a sustainable future for our planet (The Uncharted Microbial World, 2007; Hellal et al., 2023).

Beyond the borders of culturing: demystifying the "unculturable" majority

While traditional culturing techniques have been instrumental in microbial exploration, they leave a significant portion of the microbial world inaccessible. This is because many microbes cannot be readily grown under artificial conditions (National Institutes of Health (NIH); Culligan et al., 2012). The 22nd century will witness a further shift towards metagenomics, where we directly analyze the genetic blueprint of entire communities, bypassing the need for culturing. This approach, coupled with metatranscriptomics, which delves into gene expression, will provide unprecedented insights into the functional roles of diverse microbes within their complex ecosystems (Aljuraiban et al., 2023; Van Camp et al., 2023).

Current technologies only capture a fraction of microbial diversity (Vitorino and Bessa, 2018). The 22nd century will witness the development of ultra-sensitive detection methods like advanced Fluorescence-Activated Cell Sorting (FACS) and microfluidic devices that can isolate and analyze previously inaccessible microbes (Voronin et al., 2020; Daniel et al., 2022). Additionally, machine learning algorithms trained on massive datasets will be able to identify novel microbial signatures, potentially revealing entirely new branches of the microbial tree of life (Ghannam and Techtmann, 2021; Feng et al., 2023). This quest to unravel the "dark matter" of the microbial world holds immense potential for discovering novel antibiotics, bioremediation strategies (Nikolaki and Tsiamis, 2013; Escudeiro et al., 2022), and even life forms from other planets.

The 22nd century will see expeditions venture beyond familiar habitats, exploring the extreme frontiers of life, from hydrothermal vents spewing superheated fluids to the frozen depths of Antarctic ice cores (Mullineaux et al., 2018; NASA, 2024). Advanced robotics capable of operating in harsh environments will collect samples (Wong et al., 2018), while miniaturized analytical tools will enable real-time analysis, leading to the discovery of novel life forms and adaptations that push the boundaries of our understanding of life itself (Wong et al., 2018; Chellapurath et al., 2023; Soori et al., 2023). These discoveries could have implications for astrobiology, bioengineering, and our understanding of the limits of life on Earth (NASA, 2024).

The vast microbial world holds immense potential for addressing global challenges (The Microbial World, 1997). By understanding the functional roles of different microbes, we can develop solutions in diverse fields (Escalas et al., 2019; Lemke and DeSalle, 2023). Metagenomic functional analysis and protein-protein interaction networks will help us link specific genes and pathways to desired functions, leading to the development of new antibiotics, biofuels, and bioremediation strategies (de Abreu et al., 2021; Burz et al., 2023). Additionally, personalized medicine approaches tailored to individual microbiomes could revolutionize healthcare (Goetz and Schork, 2018; Wang and Wang, 2023; Goyal et al., 2023).

Microbiome of planets other than Earth, for example, Mars Microbiome, to be carried to Mars or Native Mars Microbiome?

On July 13, 2020, the virtual workshop "Microbiome for Mars" brought together leading researchers in the microbiome field to explore the relationships between the microbiome and healthcare, with a specific focus on the consequences for long-term space travel. The event delved into current microbiome research, potential future investigations, and the ways this rapidly growing field could impact astronaut health and the success of missions extending far beyond Earth. The workshop laid a foundation for the Translational Research Institute for Space Health (TRISH) to integrate microbiome considerations into its efforts to minimize the health and performance risks of human space exploration (LaPelusa et al., 2021). These space exploration by humans may cause transfer of human microbiome to any planet regardless of Mars or Moon. The delivered microbiome, either given back by the planet by reshaping genetic makeup or delivered microorganisms, may survive there for billions of years after modification in genetic makeup due to hostile conditions, as they have survived for billions of years on Earth (self-statement).

References

Aljuraiban, G. S., Alfhili, M. A., Aldhwayan, M. M., Aljazairy, E. A. and Al-Musharaf, S. (2023). Metagenomic shotgun sequencing reveals specific human gut microbiota associated with insulin resistance and body fat distribution in Saudi women. Biomolecules 13(4): 640. https://doi.org/10.3390/biom13040640.

Burz, S. D., Causevic, S., Dal Co, A., Dmitrijeva, M., Engel, P., Garrido-Sanz, D. et al. (2023). From microbiome composition to functional engineering, one step at a time. Microbiology and Molecular Biology Reviews : MMBR 87(4): e0006323. https://doi.org/10.1128/mmbr.00063-23.

Chellapurath, M., Khandelwal, P. C. and Schulz, A. K. (2023). Bioinspired robots can foster nature conservation. Frontiers in Robotics and AI 10: 1145798. doi: 10.3389/frobt.2023.1145798.

Culligan, E. P., Marchesi, J. R., Hill, C. and Sleator, R. D. (2012). Mining the human gut microbiome for novel stress resistance genes. Gut Microbes 3(4): 394–397. https://doi.org/10.4161/gmic.20984.

Culligan, E. P., Sleator, R. D., Marchesi, J. R. and Hill, C. (2014). Metagenomics and novel gene discovery: promise and potential for novel therapeutics. Virulence 5(3): 399–412. https://doi.org/10.4161/viru.27208.

Daniel, F., Kesterson, D., Lei, K., Hord, C., Patel, A., Kaffenes, A. et al. (2022). Application of microfluidics for bacterial identification. Pharmaceuticals (Basel, Switzerland) 15(12): 1531. https://doi.org/10.3390/ph15121531.

de Abreu, V. A. C., Perdigão, J. and Almeida, S. (2021). Metagenomic approaches to analyze antimicrobial resistance: an overview. Frontiers in Genetics 11: 575592. https://doi.org/10.3389/fgene.2020.575592.

Escalas, A., Hale, L., Voordeckers, J. W., Yang, Y., Firestone, M. K., Alvarez-Cohen, L. et al. (2019). Microbial functional diversity: From concepts to applications. Ecology and Evolution 9(20): 12000–12016. https://doi.org/10.1002/ece3.5670.

Escudeiro, P., Henry, C. S. and Dias, R. P. M. (2022). Functional characterization of prokaryotic dark matter: the road so far and what lies ahead. Current Research in Microbial Sciences 3: 100159. https://doi.org/10.1016/j.crmicr.2022.100159.

Feng, J., Yang, K., Liu, X., Song, M., Zhan, P., Zhang, M. et al. (2023). Machine learning: a powerful tool for identifying key microbial agents associated with specific cancer types. PeerJ 11: e16304. https://doi.org/10.7717/peerj.16304.

Ghannam, R. B. and Techtmann, S. M. (2021). Machine learning applications in microbial ecology, human microbiome studies, and environmental monitoring. Computational and Structural Biotechnology Journal 19: 1092–1107. https://doi.org/10.1016/j.csbj.2021.01.028.

Goetz, L. H. and Schork, N. J. (2018). Personalized medicine: motivation, challenges, and progress. Fertility and Sterility 109(6): 952–963. https://doi.org/10.1016/j.fertnstert.2018.05.006.

Goyal, P. A., Bankar, N. J., Mishra, V. H., Borkar, S. K. and Makade, J. G. (2023). Revolutionizing medical microbiology: how molecular and genomic approaches are changing diagnostic techniques. Cureus 15(10): e47106. https://doi.org/10.7759/cureus.47106.

Handelsman, J. (2004). Metagenomics: application of genomics to uncultured microorganisms. Microbiology and Molecular Biology Reviews : MMBR 68(4): 669–685. https://doi.org/10.1128/MMBR.68.4.669-685.2004.

Hellal, J., Lise, B., Annette, B., Aurélie, C., Giulia, C., Simon, C. et al. (2023). Unlocking secrets of microbial ecotoxicology: recent achievements and future challenges. FEMS Microbiology Ecology 99(10): fiad102. https://doi.org/10.1093/femsec/fiad102.

LaPelusa, M., Donoviel, D., Branzini, S. E., Carlson, P. E., Jr, Culler, S., Cheema, A. K. et al. (2021). Microbiome for Mars: surveying microbiome connections to healthcare with implications for long-duration human spaceflight, virtual workshop, July 13, 2020. Microbiome 9(1): 2. https://doi.org/10.1186/s40168-020-00951-5.

Lemke, M. and DeSalle, R. (2023). The Next Generation of microbial ecology and its importance in environmental sustainability. Microbial Ecology 85(3): 781–795. https://doi.org/10.1007/s00248-023-02185-y.

Mullineaux, L. S., Metaxas, A., Beaulieu, S. E., Bright, M., Gollner, S., Grupe, B. M. et al. (2018). Exploring the ecology of deep-sea hydrothermal vents in a metacommunity framework. Front Mar Sci. 5: 49. doi: 10.3389/fmars.2018.00049.

NASA. Life in the Extreme: Hydrothermal Vents. https://astrobiology.nasa.gov/news/life-in-the-extreme-hydrothermal-vents/ Assessed on 23-02-2024.

National Institutes of Health (NIH). The Human Microbiome Project. https://commonfund.nih.gov/hmp. Accessed 23 Feb 2024.

Nikolaki, S. and Tsiamis, G. (2013). Microbial diversity in the era of omic technologies. BioMed Research International 2013: 958719. https://doi.org/10.1155/2013/958719.

Soori, M., Arezoo, B. and Dastres, R. (2023). Artificial intelligence, machine learning and deep learning in advanced robotics, a review. Cognitive Robotics 3: 54–70. https://doi.org/10.1016/j.cogr.2023.04.001.

The Microbial World: Foundation of the Biosphere: This report is based on an American Academy of Microbiology colloquium held January 19–21, 1996, in Palm Coast, Florida. The colloquium was supported by the National Science Foundation, the National Oceanic and Atmospheric Administration of the U.S. Department of Commerce, the U.S. Department of Energy, and the American Society for Microbiology. Washington (DC): American Society for Microbiology; 1997. Available from: https://www.ncbi.nlm.nih.gov/books/NBK562919/ doi: 10.1128/AAMCol.19Jan.1996.

The Uncharted Microbial World: Microbes and Their Activities in the Environment: This report is based on a colloquium, sponsored by the American Academy of Microbiology, convened February 9–11, 2007, in Seattle, Washington. Washington (DC): American Society for Microbiology; 2008. Available from: https://www.ncbi.nlm.nih.gov/books/NBK562611/ doi: 10.1128/AAMCol.9Feb.2007.

Van Camp, P. J., Prasath, V. B. S., Haslam, D. B. and Porollo, A. (2023). MGS2AMR: a gene-centric mining of metagenomic sequencing data for pathogens and their antimicrobial resistance profile. Microbiome 11(1): 223. https://doi.org/10.1186/s40168-023-01674-z.

Vitorino, L. C. and Bessa, L. A. (2018). Microbial diversity: the gap between the estimated and the known. Diversity 10: 46. https://doi.org/10.3390/d10020046.

Voronin, D. V., Kozlova, A. A., Verkhovskii, R. A., Ermakov, A. V., Makarkin, M. A., Inozemtseva, O. A. et al. (2020). Detection of rare objects by flow cytometry: imaging, cell sorting, and deep learning approaches. International Journal of Molecular Sciences 21(7): 2323. https://doi.org/10.3390/ijms21072323.

Wang, R. C. and Wang, Z. (2023). Precision medicine: disease subtyping and tailored treatment. Cancers 15(15): 3837. https://doi.org/10.3390/cancers15153837.

Wong, C., Yang, E., Yan, X. T. and Gu, D. (2018). Autonomous robots for harsh environments: a holistic overview of current solutions and ongoing challenges. Systems Science & Control Engineering 6(1): 213–219. https://doi.org/10.1080/21642583.2018.1477634.

Chapter 11
A Perspective on Quantum Supercomputers and Artificial Intelligence for Microbe Hunting

Introduction

In the upcoming decades, there will be a significant shift in technical assistance and analytical methods, driven by the remarkable advancement of Machine Learning (ML), Deep Learning (DL) and Artificial Intelligence (AI) (Soori et al., 2023; Vora et al., 2023; Dwivedi et al., 2023). By the year 2030, super computers powered by artificial intelligence will enhance technical support by providing immediate troubleshooting and tailored guidance (Pew Research Center, 2023; Huwei, 2023). These instruments are designed to scrutinize extensive knowledge bases and user data in order to forecast potential issues in advance. Meanwhile, in the realm of scientific inquiry, artificial intelligence is poised to expedite the pace of new findings. Machine learning models analyze extensive datasets to uncover intricate patterns and relationships that may not be readily apparent to humans (Aldoseri et al., 2023; Mubarak et al., 2023; Elahi et al., 2023; Schmidt et al., 2019). These advancements will lead to significant progress in various areas such as drug development and materials science, marking the beginning of a new era characterized by quicker, data-informed discoveries (Bohr and Memarzadeh, 2020; Blanco-González et al., 2023). Issues arising from sample preparation, sequencing, and the application of advancing technologies may lead to inconsistencies in the data obtained currently (Gupta and Verma, 2019; Dai and Shen, 2022; Cheng et al., 2023; Akintunde et al., 2023). Even with meticulous protocols in place, slight variations are bound to occur. In the future, researchers may scrutinize the current research methodologies and potential biases found in today's data (Johnson et al., 2020; Pannucci and Wilkins, 2010; Bradley et al., 2020; Baldwin et al., 2022; Simundić, 2023). Scientific progress may reveal constraints that were previously overlooked. Subsequent research could rely on analyzing previous samples with updated methods to enhance our comprehension. To ensure the lasting significance of historical analyses, it is imperative to thoroughly document the methods used and any possible sources of error (Brown et al., 2018; Prager et al., 2019; Franco-Duarte et al., 2019; Fàbregues et al., 2021).

We are on the brink of the highly anticipated age of instantaneous DNA identification (Butler, 2015; Kaur et al., 2022; Chen et al., 2023). Enhanced by machine learning and artificial intelligence, possibly boosted by quantum supercomputers, these advancements are poised to transform the way we examine genetic material (Benschop et al., 2022; Niazi, 2023; Vilhekar and Rawekar, 2024). This extensive advancement will revolutionize areas such as forensics, personalized medicine, and ecological research, enabling us to decode the distinctive genetic information of organisms with unparalleled speed and precision, representing a significant achievement in the annals of humanity (Schork, 2019; Johnson et al., 2021; Farzan, 2024). Supercomputers in the modern era have the remarkable capacity to handle vast datasets, speeding up analysis that would be significantly slower on traditional computers (Xu et al., 2021; Verdicchio and Teijeiro Barjas, 2024). Nevertheless, substantial computational power does not inherently ensure precise outcomes. The scientific community is encountering a mounting challenge in maintaining precision due to the rising occurrence of errors and inaccuracies in analyses (Carvalho et al., 2019; Ahmed et al., 2020; Martínez-García and Hernández-Lemus, 2022). This issue arises from various potential factors. The quality of data is of utmost importance as even the most advanced computers can generate inaccurate results when provided with unreliable, insufficient, or prejudiced data ("garbage in, garbage out") (Pot et al., 2021; Bourazana et al., 2024). The intricacy of scientific models may lead to errors, especially if they are not thoroughly validated or do not accurately depict real-world phenomena. Moreover, errors made by individuals in algorithm design, result interpretation, or the management of intricate scientific supercomputing processes can result in inaccuracies. It is crucial to differentiate between authentic discoveries and artifacts resulting from errors, underscoring the importance of stringent verification procedures and a discerning approach to even the most remarkable computational results (White et al., 2016; Holleman et al., 2020; Bolger et al., 2021; Pot et al., 2021; Ahmed et al., 2020; Martínez-García and Hernández-Lemus, 2022; Bourazana et al., 2024). We are on the cusp of a new era in computing that will bring about unparalleled computational capabilities, fundamentally transforming our comprehension of the world around us. Quantum supercomputers possess the capability to conduct highly intricate calculations that surpass the capacity of classical computers. This will lead to significant advancements in microbial research, revolutionizing the realms of agriculture, medicine, and beyond (Solenov et al., 2018; Gamble, 2019; Cheng et al., 2020; Cullen et al., 2020; Mallow et al., 2022; Anand et al., 2022; Shams et al., 2023). Understanding the roles of microbes in disease, ecology, and potential applications hinges on their identification, nomenclature, and classification (Zhu et al., 2015; Rosselló-Móra and Whitman, 2019; Wei and Zhao, 2022). Novel species discovery and relationships within microbial communities (Shan et al., 2023; Zha et al., 2022) will be accelerated by quantum algorithms, streamlining these processes. The vast amounts of data produced by metagenomic analyses (Liu et al., 2022; Masenya et al., 2024; Zha et al., 2022) will pose no challenge for quantum supercomputers, enabling the identification of elusive microbial species in the dark matter. Advancements in quantum-powered bioinformatics will revolutionize the field of medicine, enabling the design of new antibiotics and targeted therapies to combat microbial resistance (Lima et al., 2019; Muteeb et al., 2023). Furthermore, pre-plannings in sustainable agriculture will be achieved through the engineering of optimized microbial communities to enhance yields and resilience (Nadarajah and Rahman, 2023; Singh et al., 2023). Enhanced environmental monitoring will allow for precise tracking of the impact of climate change on microbial ecosystems (Gámez et al., 2008; Bethwell et al., 2021; Križanović et al., 2023). To fully utilize the potential of this rapidly advancing field, it is crucial to use quantum super computing and AI tools used in microbiology.

Conclusions

Indeed, the advancement of machine learning, artificial intelligence, and the emergence of quantum supercomputers hold the potential to transform the domain of microbiology. Although the potential is vast, it is crucial to recognize the obstacles that lie ahead. Thorough sample preparation, precise sequencing methodologies, and the ongoing enhancement of technologies are essential to reduce errors and maintain the accuracy of our findings. For the upcoming scientific analysis of our work, it is crucial to maintain thorough documentation and recognize any possible constraints to ensure a well-informed interpretation within the framework of advancing knowledge. Identifying DNA in real-time with the help of AI and potentially leveraging quantum computing is on the verge of becoming a reality. This significant advancement will revolutionize various sectors, ranging from personalized medicine to environmental monitoring. The upcoming century is poised to see revolutionary findings in the field of microbiology as a result of these progressions. Despite the increasing computational power available, it is crucial for the scientific community to uphold a critical mindset that values both accuracy and innovation. In order to maximize the capabilities of these groundbreaking technologies, it is crucial to have a thorough grasp of the current hardware and software resources available for microbial detection. With the appropriate computational tools and a dedication to scientific precision, we have the ability to uncover the mysteries of the microbial realm and set the stage for a future filled with data-driven breakthroughs.

References

Ahmed, Z., Mohamed, K., Zeeshan, S. and Dong, X. (2020). Artificial intelligence with multi-functional machine learning platform development for better healthcare and precision medicine. Database : The Journal of Biological Databases and Curation 2020: baaa010. https://doi.org/10.1093/database/baaa010.

Akintunde, O., Tucker, T. and Carabetta, V. J. (2023). The evolution of next-generation sequencing technologies. ArXiv, arXiv:2305.08724v1.

Aldoseri, A., Al-Khalifa, K. N. and Hamouda, A. M. (2023). Re-thinking data strategy and integration for artificial intelligence: concepts, opportunities, and challenges. Applied Sciences 13(12): 7082. https://doi.org/10.3390/app13127082.

Anand, U., Vaishnav, A., Sharma, S. K., Sahu, J., Ahmad, S., Sunita, K. et al. (2022). Current advances and research prospects for agricultural and industrial uses of microbial strains available in world collections. The Science of the Total Environment 842: 156641. https://doi.org/10.1016/j.scitotenv.2022.156641.

Baldwin, J. R., Pingault, J. B., Schoeler, T., Sallis, H. M. and Munafò, M. R. (2022). Protecting against researcher bias in secondary data analysis: challenges and potential solutions. European Journal of Epidemiology 37(1): 1–10. https://doi.org/10.1007/s10654-021-00839-0.

Benschop, C. C. G., Slagter, M., Nagel, J. H. A., Hovers, P., Tuinman, S., Duijs, F. E. et al. (2022). Development and validation of a fast and automated DNA identification line. Forensic Science International. Genetics 60: 102738. https://doi.org/10.1016/j.fsigen.2022.102738.

Bethwell, C., Burkhard, B., Daedlow, K., Sattler, C., Reckling, M. and Zander, P. (2021). Towards an enhanced indication of provisioning ecosystem services in agro-ecosystems. Environmental Monitoring and Assessment 193(Suppl 1): 269. https://doi.org/10.1007/s10661-020-08816-y.

Bharti, R. and Grimm, D. M. (2021). Current challenges and best-practice protocols for microbiome analysis. Briefings in Bioinformatics 22(1): 178–193. https://doi.org/10.1093/bib/bbz155.

Blanco-González, A., Cabezón, A., Seco-González, A., Conde-Torres, D., Antelo-Riveiro, P., Piñeiro, Á. et al. (2023). The role of AI in drug discovery: challenges, opportunities, and strategies. Pharmaceuticals (Basel, Switzerland) 16(6): 891. https://doi.org/10.3390/ph16060891.

Bohr, A. and Memarzadeh, K. (2020). The rise of artificial intelligence in healthcare applications. Artificial Intelligence in Healthcare 25–60. https://doi.org/10.1016/B978-0-12-818438-7.00002-2.

Bolger, M. S., Osness, J. B., Gouvea, J. S. and Cooper, A. C. (2021). Supporting scientific practice through model-based inquiry: a students'-eye view of grappling with data, uncertainty, and community in a laboratory experience. CBE Life Sciences Education 20(4): ar59. https://doi.org/10.1187/cbe.21-05-0128.

Bourazana, A., Xanthopoulos, A., Briasoulis, A., Magouliotis, D., Spiliopoulos, K., Athanasiou, T. et al. (2024). Artificial intelligence in heart failure: friend or foe?. Life (Basel, Switzerland) 14(1): 145. https://doi.org/10.3390/life14010145.

Bradley, S. H., DeVito, N. J., Lloyd, K. E., Richards, G. C., Rombey, T., Wayant, C. et al. (2020). Reducing bias and improving transparency in medical research: a critical overview of the problems, progress and suggested next steps. Journal of the Royal Society of Medicine 113(11): 433–443. https://doi.org/10.1177/0141076820956799.

Brown, A. W., Kaiser, K. A. and Allison, D. B. (2018). Issues with data and analyses: Errors, underlying themes, and potential solutions. PNAS 115(11): 2563–2570. https://doi.org/10.1073/pnas.1708279115.

Butler, J. M. (2015). The future of forensic DNA analysis. Philosophical Transactions of the Royal Society of London. Series B, Biological Sciences 370(1674): 20140252. https://doi.org/10.1098/rstb.2014.0252.

Carvalho, D. V., Pereira, E. M. and Cardoso, J. S. (2019). Machine learning interpretability: a survey on methods and metrics. Electronics 8(8): 832. https://doi.org/10.3390/electronics8080832.

Chen, L., Zhou, Z. and Wang, S. Q. (2023). Process of forensic medicine in DNA identification of aged human remains. Fa yi xue za zhi 39(5): 478–486. https://doi.org/10.12116/j.issn.1004-5619.2021.511209.

Cheng, C., Fei, Z. and Xiao, P. (2023). Methods to improve the accuracy of next-generation sequencing. Frontiers in Bioengineering and Biotechnology 11: 982111. https://doi.org/10.3389/fbioe.2023.982111.

Cheng, H. P., Deumens, E., Freericks, J. K., Li, C. and Sanders, B. A. (2020). Application of quantum computing to biochemical systems: a look to the future. Frontiers in Chemistry 8: 587143. https://doi.org/10.3389/fchem.2020.587143.

Cullen, C. M., Aneja, K. K., Beyhan, S., Cho, C. E., Woloszynek, S., Convertino, M. et al. (2020). Emerging priorities for microbiome research. Frontiers in Microbiology 11: 136. https://doi.org/10.3389/fmicb.2020.00136.

Dai, X. and Shen, L. (2022). Advances and trends in omics technology development. Frontiers in Medicine 9: 911861. https://doi.org/10.3389/fmed.2022.911861.

Dwivedi, Y. K., Sharma, A., Rana, N. P., Giannakis, M., Goel, P. and Dutot, V. (2023). Evolution of artificial intelligence research in Technological Forecasting and Social Change: Research topics, trends, and future directions. Technological Forecasting and Social Change 192: 122579. https://doi.org/10.1016/j.techfore.2023.122579.

Elahi, M., Afolaranmi, S. O., Martinez Lastra, J. L. and Perez Garcia, J. A. (2023). A comprehensive literature review of the applications of AI techniques through the lifecycle of industrial equipment. Discover Artificial Intelligence 3: 43. https://doi.org/10.1007/s44163-023-00089-x.

Fàbregues, S., Escalante-Barrios, E. L., Molina-Azorin, J. F., Hong, Q. N. and Verd, J. M. (2021). Taking a critical stance towards mixed methods research: A cross-disciplinary qualitative secondary analysis of researchers' views. PloS One 16(7): e0252014. https://doi.org/10.1371/journal.pone.0252014.

Farzan, R. (2024). Artificial intelligence in immuno-genetics. Bioinformation 20(1): 29–35. https://doi.org/10.6026/973206300200029.

Franco-Duarte, R., Černáková, L., Kadam, S., Kaushik, K. S., Salehi, B., Bevilacqua, A. et al. (2019). Advances in chemical and biological methods to identify microorganisms-from past to present. Microorganisms 7(5): 130. https://doi.org/10.3390/microorganisms7050130.

Gamble, S. Quantum Computing: What It Is, Why We Want It, and How We're Trying to Get It. In: National Academy of Engineering. Frontiers of Engineering: Reports on Leading-Edge Engineering from the 2018 Symposium. Washington (DC): National

Academies Press (US); 2019 Jan 28. Available from: https://www.ncbi.nlm.nih.gov/books/NBK538701/.

Gámez, M., López, I. and Molnár, S. (2008). Monitoring environmental change in an ecosystem. Bio Systems 93(3): 211–217. https://doi.org/10.1016/j.biosystems.2008.04.012.

Gupta, N. and Verma, V. K. (2019). Next-generation sequencing and its application: empowering in public health beyond reality. Microbial Technology for the Welfare of Society 17: 313–341. https://doi.org/10.1007/978-981-13-8844-6_15.

Holleman, G. A., Hooge, I. T. C., Kemner, C. and Hessels, R. S. (2020). The 'Real-World Approach' and its problems: a critique of the term ecological validity. Frontiers in Psychology 11: 721. https://doi.org/10.3389/fpsyg.2020.00721.

Huwei. Computing 2030. https://www-file.huawei.com/-/media/corp2020/pdf/giv/industry-reports/computing_2030_en.pdf Assessed on 07-03-2023.

Johnson, J. L., Adkins, D. and Chauvin, S. (2020). A review of the quality indicators of rigor in qualitative research. American Journal of Pharmaceutical Education 84(1): 7120. https://doi.org/10.5688/ajpe7120.

Johnson, K. B., Wei, W. Q., Weeraratne, D., Frisse, M. E., Misulis, K., Rhee, K. et al. (2021). Precision medicine, AI, and the future of personalized health care. Clinical and Translational Science 14(1): 86–93. https://doi.org/10.1111/cts.12884.

Kaur, S., Kujur, M., Rawat, B., Upadhyaya, M. and Varshney, K. C. (2022). Journey of unidentified bodies towards DNA identification: A social, medico-legal and forensic perspective from New Delhi in India. Forensic Science International 341: 111470. https://doi.org/10.1016/j.forsciint.2022.111470.

Križanović, V., Grgić, K., Spišić, J. and Žagar, D. (2023). An advanced energy-efficient environmental monitoring in precision agriculture using LoRa-based wireless sensor networks. Sensors (Basel, Switzerland) 23(14): 6332. https://doi.org/10.3390/s23146332.

Lima, R., Del Fiol, F. S. and Balcão, V. M. (2019). Prospects for the use of new technologies to combat multidrug-resistant bacteria. Frontiers in Pharmacology 10: 692. https://doi.org/10.3389/fphar.2019.00692.

Liu, S., Moon, C. D., Zheng, N., Huws, S., Zhao, S. and Wang, J. (2022). Opportunities and challenges of using metagenomic data to bring uncultured microbes into cultivation. Microbiome 10(1): 76. https://doi.org/10.1186/s40168-022-01272-5.

Mallow, G. M., Hornung, A., Barajas, J. N., Rudisill, S. S., An, H. S. and Samartzis, D. (2022). Quantum computing: the future of big data and artificial intelligence in spine. Spine Surgery and Related Research 6(2): 93–98. https://doi.org/10.22603/ssrr.2021-0251.

Martínez-García, M. and Hernández-Lemus, E. (2022). Data integration challenges for machine learning in precision medicine. Frontiers in Medicine 8: 784455. https://doi.org/10.3389/fmed.2021.784455.

Masenya, K., Manganyi, M. C. and Dikobe, T. B. (2024). Exploring cereal metagenomics: unravelling microbial communities for improved food security. Microorganisms 12(3): 510. https://doi.org/10.3390/microorganisms12030510.

Mubarak, Md. H., Mimona, M. A., Islam, Md. A., Hossain, N., Zohura, F. T., Imtiaz, I. et al. (2023). Scope of machine learning in materials research—A review. Applied Surface Science Advances 18: 100523. https://doi.org/10.1016/j.apsadv.2023.100523.

Muteeb, G., Rehman, M. T., Shahwan, M. and Aatif, M. (2023). Origin of antibiotics and antibiotic resistance, and their impacts on drug development: a narrative review. Pharmaceuticals (Basel, Switzerland) 16(11): 1615. https://doi.org/10.3390/ph16111615.

Nadarajah, K. and Abdul Rahman, N. S. N. (2023). The microbial connection to sustainable agriculture. Plants (Basel, Switzerland) 12(12): 2307. https://doi.org/10.3390/plants12122307.

Niazi, S. K. (2023). The coming of age of AI/ML in drug discovery, development, clinical testing, and manufacturing: The FDA perspectives. Drug Design, Development and Therapy 17: 2691–2725. https://doi.org/10.2147/DDDT.S424991.

Pannucci, C. J. and Wilkins, E. G. (2010). Identifying and avoiding bias in research. Plastic and Reconstructive Surgery 126(2): 619–625. https://doi.org/10.1097/PRS.0b013e3181de24bc.

Pew Research Center. https://www.pewresearch.org/internet/2018/12/10/artificial-intelligence-and-the-future-of-humans/ Assessed on 07-03-2023.

Pot, M., Kieusseyan, N. and Prainsack, B. (2021). Not all biases are bad: equitable and inequitable biases in machine learning and radiology. Insights into Imaging 12(1): 13. https://doi.org/10.1186/s13244-020-00955-7.

Prager, E. M., Chambers, K. E., Plotkin, J. L., McArthur, D. L., Bandrowski, A. E., Bansal, N. et al. (2019). Improving transparency and scientific rigor in academic publishing. Brain and Behavior 9(1): e01141. https://doi.org/10.1002/brb3.1141.

Rosselló-Móra, R. and Whitman, W. B. (2019). Dialogue on the nomenclature and classification of prokaryotes. Systematic and Applied Microbiology 42(1): 5–14. https://doi.org/10.1016/j.syapm.2018.07.002.

Schmidt, J., Marques, M. R. G., Botti, S. and Marques, M. A. L. (2019). Recent advances and applications of machine learning in solid-state materials science. npj Computional Materials 5: 83. https://doi.org/10.1038/s41524-019-0221-0.

Schork, N. J. (2019). Artificial intelligence and personalized medicine. Cancer Treatment and Research 178: 265–283. https://doi.org/10.1007/978-3-030-16391-4_11.

Shams, M., Choudhari, J., Reyes, K., Prentzas, S., Gapizov, A., Shehryar, A. et al. (2023). The quantum-medical nexus: understanding the impact of quantum technologies on healthcare. Cureus 15(10): e48077. https://doi.org/10.7759/cureus.48077.

Shan, X., Goyal, A., Gregor, R. and Cordero, O. X. (2023). Annotation-free discovery of functional groups in microbial communities. Nature Ecology & Evolution 7(5): 716–724. https://doi.org/10.1038/s41559-023-02021-z.

Simundić, A. M. (2013). Bias in research. Biochem Med (Zagreb) 23(1): 12–5. doi: 10.11613/bm.2013.003. PMID: 23457761; PMCID: PMC3900086.

Singh, P., Vaishnav, A., Liu, H., Xiong, C., Singh, H. B. and Singh, B. K. (2023). Seed biopriming for sustainable agriculture and ecosystem restoration. Microbial Biotechnology 16(12): 2212–2222. https://doi.org/10.1111/1751-7915.14322.

Solenov, D., Brieler, J. and Scherrer, J. F. (2018). The potential of quantum computing and machine learning to advance clinical research and change the practice of medicine. Missouri Medicine 115(5): 463–467.

Soori, M., Arezoo, B. and Datress, R. (2023). Artificial intelligence, machine learning and deep learning in advanced robotics, a review. Cognitive Robotics 2(3): 54–70. https://doi.org/10.1016/j.cogr.2023.04.001.

Verdicchio, M. and Teijeiro Barjas, C. (2024). Introduction to high-performance computing. Methods in Molecular Biology (Clifton, N.J.) 2716: 15–29. https://doi.org/10.1007/978-1-0716-3449-3_2.

Vilhekar, R. S. and Rawekar, A. (2024). Artificial intelligence in genetics. Cureus 2024 Jan 10; 16(1): e52035. doi: 10.7759/cureus.52035. PMID: 38344556; PMCID: PMC10856672.

Vora, L. K., Gholap, A. D., Jetha, K., Thakur, R. R. S., Solanki, H. K. and Chavda, V. P. (2023). Artificial intelligence in pharmaceutical technology and drug delivery design. Pharmaceutics 15(7): 1916. https://doi.org/10.3390/pharmaceutics15071916.

Wei, W. and Zhao, Y. (2022). Phytoplasma taxonomy: nomenclature, classification, and identification. Biology 11(8): 1119. https://doi.org/10.3390/biology11081119.

White, A., Tolman, M., Thames, H. D., Withers, H. R., Mason, K. A. and Transtrum, M. K. (2016). The limitations of model-based experimental design and parameter estimation in sloppy systems. PLoS Computational Biology 12(12): e1005227. https://doi.org/10.1371/journal.pcbi.1005227.

Xu, Y., Liu, X., Cao, X., Huang, C., Liu, E., Qian, S. et al. (2021). Artificial intelligence: A powerful paradigm for scientific research. Innovation (Cambridge (Mass.)) 2(4): 100179. https://doi.org/10.1016/j.xinn.2021.100179.

Zha, Y., Chong, H., Yang, P. and Ning, K. (2022). Microbial dark matter: from discovery to applications. Genomics, Proteomics & Bioinformatics 20(5): 867–881. https://doi.org/10.1016/j.gpb.2022.02.007.

Zhu, C., Delmont, T. O., Vogel, T. M. and Bromberg, Y. (2015). Functional basis of microorganism classification. PLoS Computational Biology 11(8): e1004472. https://doi.org/10.1371/journal.pcbi.1004472.

Index

For Product Safety Concerns and Information please contact our
EU representative GPSR@taylorandfrancis.com Taylor & Francis
Verlag GmbH, Kaufingerstraße 24, 80331 München, Germany